# Sustainable Urban Development

## Topics, trends and solutions

# Sustainable Urban Development

Topics, trends and solutions

**Edited by**
**Luís Bragança**
*Civil Engineering Department, School of Engineering, University of Minho, Guimarães Portugal*

**Cristina Engel de Alvarez**
*Federal University of Espírito Santo, Vitória, Espírito Santo, Brazil*

**Luisa F Cabeza**
*Edifici CREA, Universitat de Lleida, Lleida, Spain*

**IOP** Publishing, Bristol, UK

ISBN 978-0-7503-3971-1 (ebook)
ISBN 978-0-7503-3969-8 (print)
ISBN 978-0-7503-3972-8 (myPrint)
ISBN 978-0-7503-3970-4 (mobi)

DOI 10.1088/978-0-7503-3971-1

Version: 20211201

IOP ebooks

British Library Cataloguing-in-Publication Data: A catalogue record for this book is available from the British Library.

Published by IOP Publishing, wholly owned by The Institute of Physics, London

IOP Publishing, Temple Circus, Temple Way, Bristol, BS1 6HG, UK

US Office: IOP Publishing, Inc., 190 North Independence Mall West, Suite 601, Philadelphia, PA 19106, USA

# Contents

# Foreword

Thinking the future of cities.

*Sustainable Urban Development: Topics, trends and solutions* is organized around two major problems of the moment: i) the epidemic of COVID-19, causing the death of more than 5 million people in the world, and ii) ecological imbalances, leading to climate change that threatens the lives of an even more significant number of victims and with longer-lasting consequences. The book would already contribute to the thought of this moment in the history of humanity, but its greatest originality is the focus of these two subjects under architecture and urbanism perspective.

The 16 chapters of the book cover virtually every concern in architecture and urbanism at this moment. It is worth highlighting the texts on the relationship between climate change and the city; those on new forms for urban mobility or the very concept of urbanism for the future; the analysis of the relationship between urbanism and architecture and the new concept of 'circular economy'. Several chapters deal specifically with architectural techniques and civil engineering at this time of ecological crisis and technical revolution, instigating the combination of architecture, urbanism and engineering and looking for conservation of cultural heritage and sustainability. Others search for humanization and conviviality in urban spaces. The complete set of texts presents reflections on architectural and construction techniques related to the ecological and sanitary crisis.

In addition to its broad concerns, this book provokes the reader to consider new subjects such as:

a) the need for a postmodern architecture that invents new designs that have a dialogue with Nature, avoiding technical arrogance, the indiscriminate use of air conditioning, promoting lower energy consumption and lower cost of relocation, with repair due to materials fatigue;
b) the concepts for postmodern urbanism, unifying commercial and housing spaces, reduction in the need for mobility and the size of cities, to undo the tragedies of the current 'megacities', the adaptation of the city to digital coexistence, people living in a new kind of city whose borders are no longer only defined by geographical space;
c) drawing schools to reduce the risk of epidemic contamination and incorporate the 'digital space' that will characterize broad network learning space.

This is an excellent book for its own ideas and its provocation of new ideas.

Cristovam Buarque
Emeritus Professor of Brasília University

# Acknowledgements

The editors want to thank all chapter authors, especially those belonging to the CIRES Network (Inclusive, Resilient, Efficient and Sustainable Cities), for the great effort in working completely online on this book, when we were so used to healthy, productive and fun face-to-face meetings.

The book was produced at an atypical moment, shaken by the COVID-19 pandemic, which forced us to reflect on this phenomenon in our studies, whether for buildings or the city, thus positively influencing the excellence of the contents.

We also thank Professor Dr Cristovam Buarque for dedicating his time to analyse our book, write the preface and, thus, honour us. Thank you!

We would also like to thank CYTED (Programa Iberoamericano de Ciencia y Tecnología para el Desarrollo) whose support to the CIRES Network was the fundamental tool for making this book happen. Without CYTED, the CIRES Network would not exist, and without CIRES, we would not have this book.

We also thank IOP, for the great support and the always understanding attitude towards our doubts and difficulties.

Finally, to the readers, our main target. Thank you for giving us reasons to think about the next book.

The editors

# Editors biographies

## Luís Bragança

Luís Bragança graduated in Civil Engineering at Porto University in 1982. He holds a Master degree in Thermal Engineering (1987) and a PhD in Mechanical Engineering (1995) from the same university. He is currently Professor at University of Minho (Portugal). His research interests are on the sustainability of the built environment and on evaluation tools for sustainability. He is president of the International Initiative for a Sustainable Built Environment (iiSBE). He has published more than 300 journal papers, has co-edited 6 books, and has co-authored 13 book chapters.

## Cristina Engel de Alvarez

Cristina Engel de Alvarez graduated in Architecture and Urbanism at the University of Vale do Rio dos Sinos in 1987 and holds a Master degree (1996) and PhD (2003) from the University of São Paulo. She is currently Professor Titular at the Federal University of Espírito Santo (Vitória, Brasil) and President of the Espírito Santo Research and Innovation Foundation. Her research interests are on architecture in inhospitable areas and sustainability in architecture and urbanism, and climate change. She is the coordinator of the CYTED network CIRES. She has published about 80 journal papers, has co-edited 5 books, and has co-authored 15 book chapters.

## Luisa F Cabeza

Luisa F Cabeza graduated in Chemical Engineering (1992) and Industrial Engineering (1993) at Institut Químic de Sarrià (Universitat Ramon Llull) and holds a Master Degree in Industrial Management (1995) and a PhD in Industrial Engineering (1996) from the same university. She is currently Full Professor at the University of Lleida (Lleida, Spain). Her research interests are on thermal energy storage and sustainability in buildings, processes, and cities. She is CLA of the Building chapter of IPCC AR6. She has published more than 300 journal papers, has co-edited 9 books, and has co-authored 20 book chapters.

# List of contributors

**Gilfranco Medeiros Alves**
Universidade Federal de Mato Grosso do Sul (UFMS), Cidade Universitária, Campo Grande/MS, Brazil

**Rand Askar**
ISISE, Department of Civil Engineering, University of Minho, Guimarães, Portugal

**Miguel Bandeira**
City Hall of Braga, Braga, Portugal
and
Institute of Social Sciences—Centro de Estudos de Comunicação e Sociedade (CECS), University of Minho, Braga, Portugal

**Luís Bragança**
Civil Engineering Department, School of Engineering, University of Minho, Guimarães, Portugal

**Victor Bussoloti**
Federal University of Espírito Santo, Vitória, Espírito Santo, Brazil

**Luisa F Cabeza**
Edifici CREA, Universitat de Lleida, Lleida, Spain

**Matías Cardoso**
LlactaLAB—Ciudades Sustentables, Departamento de Espacio y Población, Universidad de Cuenca, Cuenca, Ecuador

**José Pedro Carvalho**
Institute for Sustainability and Innovation in Structural Engineering (ISISE), University of Minho, Guimarães, Portugal

**Filipa Corais**
Municipality of Braga, Braga, Portugal

**Cristina Engel de Alvarez**
Federal University of Espírito Santo, Vitória, Espírito Santo, Brazil

**Anderson Azevedo Fraga**
Federal University of Espírito Santo, Vitória, Espírito Santo, Brazil

**Alvaro Valverde Garnica**
Universidad Católica Boliviana 'San Pablo', La Paz, Bolivia

**Helena Gervásio**
ISISE, Department of Civil Engineering, University of Minho, Guimarães, Portugal

**Carlos H Godoy**
CEA—Centro de Energía y Ambiente, Universidad Tecnológica Nacional, Argentina

**Begoña Guirao**
Universidad Politécnica de Madrid, Madrid, Spain

**Fermín Rodríguez Gutiérrez**
Departament of Geography, Universidad de Oviedo, Mieres, Asturias, Spain

**M Augusta Hermida**
Facultad de Arquitectura y Urbanismo, Universidad de Cuenca, Cuenca, Ecuador

**Gustavo Kamino**
Civil Engineering Department, School of Engineering, University of Minho, Guimarães, Portugal

**Nils Larsson**
International Initiative for a Sustainable Built Environment (www.iisbe.org), Ottawa, Ontario, Canada

**Karina Trevisan Latosinski**
Universidade Federal de Mato Grosso do Sul (UFMS), Cidade Universitária, Campo Grande/MS, Brazil

**Ricardo Mateus**
Institute for Sustainability and Innovation in Structural Engineering (ISISE), University of Minho, Guimarães, Portugal

**Norberto S Odobez**
CEA—Centro de Energía y Ambiente, Universidad Tecnológica Nacional, Argentina

**Daniel Orellana**
Facultad de Ciencias Agropecuarias, University of Cuenca, Cuenca, Ecuador

**Fátima Pereira**
Municipality of Braga, Braga, Portugal

**Ilaria Pigliautile**
CIRIAF—Interuniversity Research Center, Department of Engineering, University of Perugia, Perugia, Italy

**João Onofre Pereira Pinto**
Universidade Federal de Mato Grosso do Sul (UFMS), Cidade Universitária, Campo Grande/MS, Brazil

**Anna Laura Pisello**
CIRIAF—Interuniversity Research Center, Department of Engineering, University of Perugia, Perugia, Italy

**Kamel Mohamed Rahla**
Institute for Sustainability and Innovation in Structural Engineering (ISISE), University of Minho, Guimarães, Portugal

**Edna Aparecida Nico Rodrigues**
Federal University of Espírito Santo, Vitória, Espírito Santo, Brazil

**Ricardo Rodrigues**
Faculdade de Arquitectura, Centro de Estudos de Arquitectura e Urbanismo, Universidade do Porto, Porto, Portugal

**Vladimir Fernando Stello**
Instituto do Patrimônio Histórico e Artístico Nacional—IPHAN, Escritório Técnico de Laguna, Laguna, Brazil
Escritório Técnico de Laguna

**Juliana Couto Trujillo**
Universidade Federal de Mato Grosso do Sul (UFMS), Cidade Universitária, Campo Grande/MS, Brazil

**Maria Matilde Villegas**
Universidade do Sul de Santa Catarina—UNISUL, Tubarão, Brazil

**Andrea Naguissa Yuba**
Universidade Federal de Mato Grosso do Sul (UFMS), Cidade Universitária, Campo Grande/MS, Brazil

# Introduction

This book was conceived in 2020, at a time when the COVID-19 pandemic deeply shook the certainties of knowledge produced by science and technology. It seemed that all the theses produced so far had lost their value or were being challenged by a jury, not of experts, but of common people who suffered and wept for their countless dead.

Science was forced to seek new and faster paths. We were confronted in our research about the built environment by sometimes simple questions: what to do with the studies of mass transport? And what about the previously unquestionable incentive to build urban spaces that sought to create opportunities for large events and the consequent agglomeration of people? From one moment to another, the utopias so passionately pursued by researchers and professionals from various sectors seemed to have lost their meaning. Is the intelligence of cities, in fact, a question of improving the quality of life for all? And sustainability—will it be worth the effort to continue fighting for it?

2021 has arrived, and if, on the one hand, it filled us with the hope of a way to face the pandemic, on the other hand, it has brought us the real impacts—and no longer those seen as possible occurrences in simulations—of climate change. We witness planetary imbalance through floods in places that have never seen them before, terrible droughts, and even snow in tropical countries.

We found that we will have to reinvent ourselves, but on the other hand, the knowledge produced so far can and should be used to formulate policies that lead to a search for more humane and equitable, safer, and healthier cities.

Keeping all these new challenges in mind, this book is structured in two parts: the first is related to urban issues and the second to buildings.

The authors come from different realities from 11 different countries: Algeria, Argentina, Bolivia, Brazil, Canada, Colombia, Ecuador, Spain, Italy, Syria, and Portugal, whose approaches were directed to themes previously selected as priorities, such as climate change, energy, historical heritage, urban mobility, social aspects, and materials, among others.

In summary, this book offers two specific but deeply correlated approaches: the scale of the city and the building scale. The content covered begins by dealing with issues related to climate change, its effects, and the need to seek mitigation and face the severe consequences already experienced in most countries (chapter 1). The urban issues continue to be discussed, now focusing on the participatory planning process considering the dynamics of urban agglomerations and the need for effective flexibility according to the priorities of each time, as amply demonstrated in recent years (chapter 2). Next, urban mobility and urban open spaces are approached from different perspectives: in a broader view, in which we consider new concepts for the design of cities (chapter 3); the use of information and communications technology in urban mobility and the need to expand non-motorised transport, especially with the advent of COVID-19 (chapter 4); a case study from the concept of 'urban acupuncture' (chapter 5); and adding value to public spaces considering tactical urbanism as the basis of decision-making (chapter 6).

At a time of great transition and questions related to the need for changes in behaviour and the directions of public policies, it is inevitable that aspects related to social issues and the search for strategies that solve the basic problems of the most vulnerable communities will be discussed (chapter 7). We also consider the values related to historical heritage and the new challenges in the search for the preservation of the cultural identity of a people, a theme of fundamental importance, particularly in the light of the shock caused to convictions about what should be considered a priority in urban planning (chapters 8 and 9).

The pandemic also accelerated a discussion that was already taking place regarding a change in the use of buildings. The increasing use of the home office and distance learning, for example, in addition to reducing the need for spaces in commercial and school buildings, also brought to light the new uses of residential compartments. Thus, issues such as the flexibility of architectural solutions must be considered in the design of buildings (chapter 10), which, of course, must be accompanied by new procedures in the building design process, through instruments that help in more efficient decision-making (chapter 11).

When thinking about the architectural object, that is, at the scale of a building, another emerging theme is energy and the search for efficiency in solutions related to both materials and techniques and the need to change the energy matrix. The desired model is the autonomous-energy building, which can produce the equivalent of (or close to) its consumed energy (chapter 12), or even the consideration of the slightly larger scale of a set of buildings or similar strategies (chapter 13). However, the basis of the energy issue is also the constant increase in the consumption of construction materials, and the analysis should be extended to the so-called embodied energy, whose study is directly correlated to $CO_2$ emission (which has an undeniable influence on climate change), and to the effective application of the concept of the circular economy (chapter 14). In particular, when it comes to the circular economy, although the need to reduce demolitions (with the consequent increase in waste generation) seems obvious, the removal of old buildings for the construction of new ones is still a reality in most urban centres. Thus, the existing real-estate use still requires feasibility studies and specific measures that can induce incentive policies for construction entrepreneurs (chapter 15).

This book ends by closing the cycle of reflections at the point where it started, returning to the two main issues that have shaken the lives of all citizens: the COVID-19 pandemic and climate change (chapter 16). More than simply facing these issues, which definitely deeply affect people's quality of life, such events and their consequences can be seen as an important historical moment that generates reflections on our society's current lifestyle and values and the opportunities that arise from a time of crisis.

The editors wish all readers good and fruitful reading.

Luís Bragança
Cristina Engel de Alvarez
Luisa F Cabeza
(*Editors*)

**IOP** Publishing

Sustainable Urban Development
Topics, trends and solutions
**Luís Bragança, Cristina Engel de Alvarez and Luisa F Cabeza**

# Chapter 1

## The relationship between climate change and the city

**Cristina Engel de Alvarez, Gustavo Kamino, Victor Bussoloti and Luís Bragança**

The global warming caused by anthropic actions, especially those related to the emission of greenhouse gases (GHGs), has shown itself to be unquestionable. Most people currently live in cities, and this massive urbanisation raises the impacts of climate change to a new level and makes the preservation of the quality of cities and their resilience increasingly important. According to the studies performed in this field, the effects related to climate change will be especially felt in terms of temperature, rainfall, droughts, floods and runoff from rivers, oceans, cyclones, and extreme events. However, the effects on each city will vary according to the kind of impact it will suffer, the level of the city's development, and how prepared it is for the consequences. This is because climate change can result in a ripple effect across urban infrastructure systems which are interdependent on each other, such as the water, sanitation, energy, and transportation sectors. The vulnerability of these sectors to climate change depends mainly on their degree of development, resilience, and ability to adapt. Therefore, the transition to low-carbon cities is critical to the sustainability movement. The relation between urbanisation and GHG emissions is complex. With the rise of the COVID-19 pandemic, the idea of recovery from the economic crisis evolved into a concept more suited to fighting climate change and its consequences. Thus, the concept was established of considering the need to reduce or eliminate heavily carbon-producing economic activities. Low- and middle-income cities face additional challenges due to increasing urbanisation, social and economic inequality, migration, and informal settlement. In addition, many of these cities are in areas that are more vulnerable to climate change impacts, which make the adaptation and mitigation efforts even more crucial. Climate actions in such places must also consider inclusiveness and efficiency as key aspects for the development and improvement of people's lives, and a path towards sustainability.

doi:10.1088/978-0-7503-3971-1ch1  

## 1.1 Introduction

Since the Fourth Assessment Report of the Intergovernmental Panel on Climate Change—IPCC (2007), issues related to climate change have passed the stage of debate and have become unequivocal truths. The global warming caused by anthropic actions, especially those associated with GHG emissions, has shown itself to be unquestionable. This global change caused by humans is affecting the whole of the environment and the biosphere. Over the decades, GHG emissions have altered climate standards and have resulted in worldwide warming. Oceans, forests, glaciers, deserts, and all the other natural environments are changing, so that life must adapt in order to survive. Climate change is already provoking humans to change how they live and consequently to consider how buildings and cities should be designed.

Studies related to this theme have been developed since the middle of the last century, but the discussions have become more intense, and with the involvement of society in general since the 1980s, easily noticeable, due to the governmental and non-governmental actions aimed, in particular, at the search for a reduction of environmental loads. In this respect, the various agreements established by international meetings and conferences with the participation of leaders from the most important nations stand out. However, although such events and the commitments made by participating nations have been widely celebrated and publicised by the media, the desired effects have not been obtained, especially in countries whose economies are based on industrial activity.

The main agreement currently in force is the so-called Paris Agreement, the international treaty on climate change adopted by 196 countries at COP 21 (Conference of the Parties) in Paris on December 12, 2015, which came into force on November 4, 2016 (UN-Climate Change 2019). It aims to limit global warming to less than 2 °C, but preferably to 1.5 °C, compared to the temperatures recorded before the Industrial Revolution in the 18th century, when the population growth in cities started.

However, the planet is currently far from meeting the 1.5 °C or 2 °C targets of the Paris Agreement. In recent years, heat records have been broken; 2019 was the second warmest year on record and the decade from 2010 to 2019 was the warmest on record. Since the 1980s, every decade has been warmer than any previous decade since 1850 (UN-News 2020). Additionally, new researchers are revisiting methods and datasets and updating the warming estimate in increments of at least 1 °C (IPCC 2021).

As a result of the global pandemic (COVID-19), due to travel bans and the economic slowdown, it is estimated that there may have been a drop of approximately 6% in GHG emissions, however, there is no doubt that this improvement was only temporary. There has been no halt to ongoing climate change, and, as the economy starts to recover, emissions are also returning to previous high levels (UN 2020).

Most people currently live in cities. In 2018, about 55% (4.22 billion) of the world population lived in urban areas, and according to projections, this number will be 68% (6.88 billion) by 2050. This represents 2.6 billion more people living in cities, an increase of 63% in just 30 years, mainly located in developing and less developed countries. This massive urbanisation will raise the impacts of climate change to a new level and make the preservation of the quality of cities and their resilience increasingly important.

More and more data and findings show that climate changes, especially warming, are inducers of extreme events, as shown by the increase in their frequency and magnitude of occurrence. Such warming intensifies the continuity of, and accelerates the increase in, the average global temperature. It is estimated that among the total disasters that have occurred in the last 20 years, 91% have been climate-related disasters.

With the increase of urbanisation and climate change, in addition to the interference with air quality, aspects related to water availability and quality, land use, and the production and management of waste, it should be noted that extreme events tend to have more significant consequences for the most vulnerable populations, increasing further social inequities.

The health of the urban population can be strongly impacted, while extreme heat can cause reactions in the human body ranging from simple cramps to even exhaustion and heat stroke. These reactions, which often affect the body's ability to maintain its normal temperature, occur more frequently in vulnerable groups, such as children, the elderly, and people with pre-existing illnesses. Moreover, most of the impact factors are out of man's control, such as their geographic location, altitude, proximity to the ocean, water accessibility, etc. Nonetheless, these impacts cause profound effects on human life, city planning, infrastructure, and public services, which require immediate attention and action from governments and society. For example, in 2019, high temperatures caused more than 100 deaths in Japan and 1462 deaths in France (UN-News 2020).

New challenges arise every day. Globalisation has brought development to areas otherwise forgotten, rapidly connecting people and places never seen before. Likewise, it presents unthinkable challenges that a world in global change must be prepared to face.

Nevertheless, only in the first decade of this century have issues related to the need for cities to adapt become part of the agenda for discussion in international climate-related conferences. As a result, cities have become the main contributors to the intensification of climate change problems and also the recipients of its effects (Alvarez and Bragança 2018).

Urban resilience, which is the ability of a city to resist, absorb, adapt, and recover from exposure to threats, is becoming a concept that is integrated into public policies, whether local or global. In parallel, it is necessary to adopt measures that effectively reduce GHG emissions, slowing the growth rate of climate change, in particular, global warming.

It is essential to consider studies related to the expected direct and indirect impacts, given the +1.5 °C scenario established by the Paris Agreement (and also the IPCC +2.2 °C scenario), and how these environmental impacts will affect cities' infrastructures, housing, transport, health, ecosystems, water, energy, and solid waste management. The questions that should be asked are how cities can respond in order to adapt and to mitigate effects, and how they can do it by being more inclusive, resilient, and efficient.

## 1.2 Global warming effects

GHG emissions due to human activities have been increasing since pre-industrial times, driven by population growth and economic development. These carbon dioxide, methane, and nitrous oxide emissions concentrate in the atmosphere and have reached an unprecedented level. Studies show that they have been the main cause of warming since the mid-20th century (IPCC 2014).

Also, according to the IPCC reports, the rates of global carbon dioxide ($CO_2$) emission have increased by almost 50% since 1990. In addition, the GHG emissions have grown more rapidly in the period between 2000 and 2010 than in each of the previous three decades (UN 2020).

Anthropogenic effects are responsible for approximately 1.0 °C of global warming above pre-industrial levels. Studies show a wide range of possible future global mean temperature increases, from 1.5 °C to 4.8 °C from the pre-industrial era to the end of the 21st century, based on different scenarios for GHG emission, atmospheric concentration, air-pollutant emission, and land use. The IPCC (2014) defends a rigorous mitigation procedure, limiting emissions to achieve the 1.5 °C scenario.

Five emission scenarios called SSPs—Shared Socioeconomic Pathways—were simulated by the 6th Assessment Report of the IPCC. First, the scenarios consider very low and low GHG emissions, with $CO_2$ emissions declining to net zero by around 2050, followed by varying levels of net negative $CO_2$ emissions (SSP 1–1.9 and SSP 1–2.6). Next followed an intermediate scenario (SSP 2–4.5), with $CO_2$ emissions remaining at around current levels until 2050. Finally, the report described the most critical scenarios with high and very high GHG emissions (SSP 3–7.0 and SSP 5–8.5), and $CO_2$ emissions doubling from current levels by 2100 and 2050, respectively. In the most adverse scenario, in some regions of the planet, especially in the far North, a temperature increase of up to 11 °C would be observed, in addition to extremes of heat noticed mainly in Latin America for more than 160 days a year.

The IPCC report further states that it is still possible, through technological measures, changes in behaviour, and institutional actions to limit the increase in the average global temperature to 2 °C above pre-industrial levels. However, the measures used to reach the increasingly unlikely scenario of a maximum temperature increase of up to 1.5 °C need to be urgent. In this sense, the establishment of public policies for effective action is of fundamental importance in changing the scenery and responsibility for this necessary change cannot be transferred to the individual or to an alleged collective consciousness.

Observations and measurements throughout the last decades have shown changes in climates and weather. Observed temperatures show that in the last three decades, Earth's surface has become successively warmer. In the Northern hemisphere, the period from 1983 to 2012 was probably the warmest of the last 1400 years. The ocean's accumulated energy has increased drastically, corresponding to more than

90% of the energy stored in the oceans compared to only 1% in the atmosphere between 1971 and 2010. In addition, the excessive $CO_2$ uptake by oceans has resulted in acidification and the pH of surface water has decreased by 0.1; precipitation measured in the mid-latitude areas of the Northern hemisphere has increased. It should also be remembered that high temperatures and intense solar radiation enhance the photochemical reactions responsible for emitting gases and polluting particles into the atmosphere.

Climate change impacts can already be observed in some natural systems. For example, hydrological systems are being altered by the change of precipitation and snow/ice melt; faunal observations show species that have changed their interactions, geographic distribution, seasonal activities, migration patterns, and population due to climate change; and the Greenland and Antarctic ice sheets have been losing more and more mass over the period from 1992 to 2011 (figure 1.1).

According to studies in this field, the effects related to climate change will be especially felt in terms of temperatures, rainfall, droughts, floods, and runoff from rivers, oceans, cyclones, and extreme events.

According to the IPCC (2018), continental ***temperature*** changes are already widely known and are mainly detectable in North America, Europe, and Australia, with a global increase in the frequency and intensity of hot days and nights and a decrease in the intensity and frequency of cold days and nights.

Projections of future temperatures are linked to the mitigation scenarios predicted by the IPCC (2018). Considering the 2046–2065 timeframe, a strict mitigation scenario could lead to a global mean temperature increase of 1.0 °C, with a likely

**Figure 1.1.** The Antarctic environment, especially in coastal areas, tends to be one of the first environments impacted by climate change, both due to the melting of glaciers and the rise in sea level.

range of 0.4 to 1.6 °C; intermediate scenarios could lead to a mean increase of 1.4 °C, with a likely range of 0.9 °C to 2 °C, and a high GHG emission scenario could lead to a 2.0 °C mean increase, with a likely range of 1.4 °C to 2.6 °C (IPCC 2014).

There is an expected increase in the average temperature variability for tropical regions, which are more susceptible to heatwaves and thermal extremes, with a progressive number of warmer days in the pessimistic future scenarios. This thermal tendency in tropical cities aggravates the imbalance between social, economic, and environmental aspects in developing countries.

The year 2021 may have been one of the hottest years, with forest fires, droughts, storms, and intensified melting of glaciers. Recently, torrential rains in Belgium and Germany have flooded several rivers, destroying thousands of buildings and killing many people. In the Chinese city of Zhengzhou, the equivalent of a whole year's rain fell in just three days. In Canada, in June, the maximum temperature record was reached (49.6 °C) and in July, Finland experienced its hottest night ever. Wildfires in Australia, California, and the Mediterranean basin are increasingly longer and more intense; they devastate larger areas, causing health problems due to the smoke, deaths, and massive damage. All of this happened with an Earth warmed by just 1.1 °C–1.3 °C. The climatic collapse signs are becoming more visible each year. With the emissions estimates for the year 2030, the planet is on its way to a 3.2 °C rise in temperature this century, even if the promises of governments under the Paris Agreement (known as Nationally Determined Contributions (NDCs)) are fully implemented.

In the models and simulations carried out so far, regardless of the database used, the results invariably indicate that the distribution of the global warming will be highly unequal. The Arctic region will warm more and faster than other regions. High latitudes in the Northern hemisphere will concentrate great temperature changes. Globally, land areas will capture more heat than oceans. Tropical regions are likely to have a smaller increase in mean temperature, but as they approach threshold temperatures, such an increase can lead to great ecosystem impacts (IPCC 2018).

As already mentioned, this temperature rise can lead to adverse reactions in humans, such as cramps, exhaustion, and even fatal medical emergencies, if not quickly treated. In addition, pre-existing cardiorespiratory pathologies can suffer sudden worsening.

The aspects related to precipitation also deserve to be highlighted, since changes in precipitation will not be uniform. The projections for precipitation are not as accurate as those for temperatures, but it is expected that more areas will experience increases in frequency and intensity than the areas with a rain reduction.

Projections show that North America, North Europe, parts of Asia, and Africa will have equal or high precipitation levels. Meanwhile, North and South Africa, South and Central America, Southern Europe, and East Asia may have less precipitation (IPCC 2018).

Research in Africa found that the average ***precipitation*** may change significantly due to fewer rainy days that are compensated for by heavier rains and less rainfall in early summer and more in late summer. This scenario can be extremely dangerous, especially for the most socially vulnerable communities.

In Europe, changes in precipitation may also occur. Studies show that while the average precipitation is expected to increase in Central and Northern Europe in the winter, this happens only over Northern Europe in the summer, while it decreases in Central and Southern Europe.

Although confidence in the projected precipitation changes is not as high as for the temperature changes, the responses in the model simulation are robust and predictably linear. Significant differences in rain distribution were not found between the 1.5 °C and the 2 °C global warming scenarios. However, in the 2 °C scenario, there will be more heavy precipitation (IPCC 2018). It should also be noted that many platforms that offer simulation programs obtain quite different results, even when the input data are similar. Thus, special attention must be paid to the reliability of these sites.

In scenarios for temperature changes between 3.2 °C and 5.4 °C, an increase up to 50% in the average rainfall is expected in the extreme north of the planet at the end of the century. Regions such as the Northeast of South America, Central America and Southern Africa will, in turn, suffer from a reduction in rainfall of up to 30%.

There is low confidence in the **drought** and dryness trends on a global scale, but more confidence in regional observations. The Mediterranean and Western African regions may experience increases in dryness, and Central North America and Northwest Australia will see drought decrease. Modelling studies also project an increase in drought years in the Mediterranean region. An increase in drought and dryness, or a reduction in precipitation is foreseen in some regions as a result of considering the 1.5 °C and 2 °C global warming levels. However, there are many variability issues that affect the precision, except for the Mediterranean region, which has more data. Analysis that considered evapotranspiration has identified hotspots of drying in the Mediterranean region, Northeastern Brazil, and Southern Africa.

The probability of dryness increases with global warming, given that the 2 °C or more global warming scenarios tend to further reduce water availability in regions such as the Mediterranean and Southern Africa. However, according to the IPCC (2018) studies, constraining global warming to 1.5 °C will avoid the most extreme risks of change.

Additionally, the ingestion of, or contact with, polluted water can result in many types of disease. Pathogens that cause illness may reproduce faster due to climate change effects. Floods also potentiate the transport of too many pollutants and sediments into water and food supplies. The scarcity in quantity and the reduced quality of food and water, due to droughts, can impact human nutrition and health.

The likely changes in precipitation, associated with temperature changes that affect the water cycle, will also have consequences for cities located close to rivers, deltas, and islands. Studies show that in some regions there will be an increase in ***flood*** frequency, extreme ***streamflow***, mudslides, and flood hazards (IPCC 2018). Extreme hydrological events will occur, and the annual global exposure to floods is expected to increase by between four and fourteen times over the 21st century compared to the 20th century.

The global projected changes in average annual streamflow are smaller for the 1.5 °C scenario than for 2 °C of global warming. A study of 21 of the most important river basins showed that most of them will experience an increase in runoff; the exceptions are the Guadiana, the Danube, the Amazon, and the Orange basin, for which the projections forecast a decreased flow.

There is a lot of confidence that global warming will lead to an expansion of areas of flooding and increases in runoff, thus incrementing hazards. Once again, the most socially vulnerable riverside populations will be the most affected, given the impossibility of relocating to safer places.

Coastal and low-lying areas have great vulnerability to storms and heavy precipitation. The impact of these depends heavily on their specific characteristics, but they can directly cause deaths and injuries and indirectly cause effects such as the contamination of soil and water, along with damage to respiratory and mental health. Mental health impacts are particularly common and strong, as they frequently come after the stress of evacuation, property damage, economic loss, and death traumas.

It is essential to realise that climate change can provoke an imbalance in ecosystems and directly influence the reproduction of species. For example, malaria and dengue are mosquito-borne diseases that strike many developing countries' populations and directly correlate with warming trends. The incidence of dengue virus increased in 2019 due to higher temperatures, which facilitated mosquitoes' transmission of the disease (UN-News 2020). Likewise, a tick, a vector for Lyme disease, has expanded from the US to Canada due to an increase in temperature.

Regarding the ***oceans***, although it is estimated that they will continue to warm up, this increase in temperature will occur at rates lower than the rate of increase in the global temperature. As a result, the frequency of occurrence of marine heatwaves will grow. Sea ice levels between 1979 and 2012 showed a decrease of 3.5% to 4.1%; projections show that this process will continue in all scenarios, and at least once a century, the Arctic summer will be ice-free (IPCC 2018).

Ocean warming is expected to be greater in tropical and northern subtropical regions and more significant in the southern ocean. As Atlantic Meridional Overturning Circulation (AMOC) becomes weaker, global warming will increase (IPCC 2014).

The global mean sea level will continue to rise in all scenarios, at rates greater than 2 mm $yr^{-1}$. The sea-level change will not be uniform, although it will rise over more than 95% of the ocean area, and 70% of coastlines worldwide. The sea-level rise is related to $CO_2$ emissions, and the sooner its reduction occurs, the better (IPCC 2014).

In the absence of anthropogenic climate change, sea levels on the planet would have risen by less than half of the amount seen during the 20th century and could even have fallen. However, the rate of sea-level rise in the past two decades is almost double that of the last century, and each year it is accelerating slightly. From 1901 to 2010, the average sea level on the planet increased by 19 cm because the oceans expanded due to the warming and melting of ice. In the Arctic, sea ice has shrunk every decade since 1979, by 1.07 million $km^2$ per decade (UN 2020).

Ocean chemistry is also changing. The increase in the water surface temperature impacts water oxygenation, which has reduced since 1960, affecting ocean life. As a result of increasing the $CO_2$ in the atmosphere, the ocean is also increasing in $CO_2$ content and becoming more acid. Since the beginning of the Industrial Revolution, the acidity of the surface ocean waters has increased by about 30%, as its pH has fallen by 0.1 units. Ocean salinity is also changing with hotter temperatures, becoming less salty in the Arctic (due to ice melting) and saltier in other areas (due to increased evaporation), potentially affecting large-scale changes in water movement.

The fact that the oceans are becoming more acidic affects many species, such as oysters—due to their shells—and corals, which may even start to dissolve if the pH gets too low. Researchers have even discovered severe levels of dissolution of the shells of some species in the Southern Ocean, which surrounds Antarctica. In addition, the ability of some fish to detect predators is reduced in more acidic waters. However, algae and seagrass can benefit from the higher $CO_2$ conditions in the ocean, as they require it for photosynthesis. Some studies have examined whether cultivating seaweed could help to slow the ocean's acidification, as seaweed is mainly responsible for the oxygenation of the planet.

With the modification of temperatures, precipitation, and water cycles, there is a natural tendency for changes to also happen in the wind system. However, there is low confidence in the changes in ***cyclones*** observed since the pre-industrial age. Projections show an increase in heavy precipitation associated with tropical cyclones (IPCC 2018).

Although the simulation results point to a global decrease in the number of cyclones due to the increase in temperatures, there is a tendency for the frequency of occurrence of the highest-intensity cyclones to increase (IPCC 2018).

In 2019, ***extreme weather events*** occurred, some on an unprecedented scale, in different parts of the world. The monsoon season featured above-average long-term rains in India, Nepal, Bangladesh, and Myanmar, and flooding in the region killed 2200 people. In 2019, more tropical cyclones occurred than the average: 72 in the northern hemisphere and 27 in the southern hemisphere. Cities on the east coast of Africa, Japan, and the Bahamas suffered from destruction and/or flooding. Part of South America was hit by floods in January, while Australia had its driest year on record. South Africa, Central America, and parts of South America had abnormally low rainfall rates. In the United States, floods generated losses of US $ 20 billion (UN-News 2020).

## 1.3 City infrastructure and services impacts

Cities are central areas for human settlement and economic activities; projections show that by 2050, two-thirds of the world's population will live in cities. Most of this urbanisation will occur in Asia and Africa. Regarding GHG emissions, cities are part of both the problem and the solution. Approximately 70% of global anthropogenic GHG emissions are emitted into the urban environment, but the per-capita energy consumption of urban residents in developed countries is lower than that of rural residents (UN-Habitat 2016).

Climate change impacts will have very extensive consequences for urban areas on every continent. Cities have unique climate risks, such as urban heat islands, impervious surfaces that exacerbate flooding, and coastal developments which are threatened by sea-level rise. In addition, urban areas' characteristics, such as their heat-absorbing materials, reduced evaporative cooling, lack of vegetation, and produced waste can lead to excessive heat compared to other areas.

Existing risks will be amplified, and new threats will be created, especially for people in disadvantaged and low-income areas. Climate extreme events, such as heatwaves, droughts, heavy precipitation, and floods will be exacerbated in frequency and severity, leading to casualties and damage, leading to calls for measures to increase resilience to mitigate such impacts (figure 1.2).

The effects of the global change in each city vary not only according to the kind of impact it will suffer but also according to the level of development of the city and how prepared it will be for the impacts of global change. Therefore, climate change risks need a transdisciplinary approach to the climate system, natural and managed ecosystems, human systems, and responses (IPCC 2018).

This is because climate change can result in a ripple effect across urban infrastructure systems, which are interdependent on each other, such as the water, sanitation, energy, and transportation sectors. The vulnerability of these systems to climate changes depends on their degree of development, resilience, and ability to adapt. Climatic variations can aggravate the problems, especially in developing countries, where a significant part of the population still lacks basic sanitation services, there are serious traffic problems, and energy security is still an issue for discussion (PBMC 2016).

Among the various parts that make up an urban community, the ***housing*** sector can be considered the heart of the city, accounting for a large part of the land area, the energy consumption, and the population's wellbeing. Most of the global energy demand for buildings is generated by developed countries. Building renovations are opportunities to apply long-term energy-saving solutions using better house designs

**Figure 1.2.** View of Vitoria (Brazil), whose social contrasts are representative of most cities located in tropical regions, especially Latin America.

and to retrofit to improve energy consumption and reduce GHG emissions. Renovated spaces can also improve quality of life by providing better public areas and local job creation (Jean-Baptiste *et al* 2018).

Low-income and informal settlements are especially vulnerable to climate change threats due to their geophysical locations, land availability, poor infrastructure, low house quality, and limited support in disasters. Table 1.1 shows the possible impacts of global warming changes on urban areas and how they can impact poor areas and informal settlements. Among those impacts, many are related to health, like the increased rate of disease, access to water, and life expectancy; others are related to housing, such as durability, and still more are related to the economy, like employment and ownership.

Some impacts listed in table 1.1 have already been noticed due to climate change. In 2019, about 6.7 million people on the planet had been displaced from their homes due to natural disasters—such as storms, floods, and devastating cyclones (UN-News 2020).

Low- and middle-income countries still face a high demand for housing; an estimated 500 million people will be in need of a home by 2050 (Jean-Baptiste *et al* 2018). This represents a big challenge and an opportunity to build cost-efficient, low-carbon, and resilient dwellings while improving access and reducing climate-related risks. In addition, it should also be considered that, in the future, the COVID-19 pandemic may change both population growth and the relationship between the user and the environment, probably making people more dependent on air conditioners and more energy for household and office activities.

Associated with the issue of housing, aspects related to ***urban mobility*** should be discussed. A great proportion of GHG emissions are from urban transportation systems. Cities are responsible for 70% of $CO_2$ emissions, and the transport sector accounts for 30% of the total. Urban transportation accounts for 40% of total transport emissions, with a growth of about 2% to 3% annually. The majority of the GHG stock belongs to developed countries, but 90% of the growth corresponds to lower-income countries (Mehrotra and Zusman 2018).

Adequate planning for public transport systems can provide several mobility options, increase public safety, reduce vehicle travel distances, reduce air pollution, reduce energy consumption, conserve natural resources and open spaces, reduce infrastructure costs, and contribute to more affordable housing. Areas of relatively high density are necessary to generate greater efficiency in the use of public transport. In general, density is one of the factors that affects energy expenditure and GHG emissions. Dealing with these issues requires an ongoing analysis of urban processes (UN-Habitat 2011).

Regarding air pollution, it should be noted that one in seven deaths around the world is linked to air pollution. Furthermore, climate variations can potentially change emissions, transport, dilution, chemical patterns, and the deposition of air pollutants. For example, the increase in wildfires leads to particle emissions and respiratory disease.

Urban transportation is vital to city living and there are many interdependencies between transport and other areas. Climate-related events can cause damage to the

**Table 1.1.** Climate change impacts on urban and informal settlements. Source: Adapted from Jean-Baptiste *et al* (2018).

| Climate change effect | Possible impacts on urban areas | Possible impacts on informal settlement areas |
|---|---|---|
| Temperature | Increased energy demand<br>Worsening of air quality<br>Exaggerated by urban heat islands | Disease, morbidity, and mortality<br>Health and life expectancy<br>Access to water and sanitation |
| Precipitation | Increased risk of flooding<br>Increased risk of landslides<br>Distress-driven migration from rural areas<br>Interruption of food supply networks | Life expectancy<br>Access to water and sanitation<br>Access to adequate living space<br>Durability of dwellings<br>Security of ownership |
| Sea-level rise | Coastal flooding<br>Reduced income from agriculture and tourism<br>Salinisation of water sources | Durability of dwellings<br>Access to water and sanitation<br>Security of ownership<br>Labour and employment |
| Extreme rainfall | More intense flooding<br>Higher risk of landslide<br>Disruption to livelihoods and city economies<br>Damage to homes and businesses | Durability of dwellings<br>Disease, morbidity, and mortality<br>Life expectancy<br>Access to adequate living space<br>Security of ownership<br>Labour and employment |
| Drought | Water shortages<br>Higher food prices<br>Disruption of hydroelectricity<br>Distress-driven migration from rural areas | Access to water and sanitation<br>Security of ownership<br>Diseases, morbidity, and mortality |
| Heat or cold waves | Short-term increase in energy demands for heating/cooling | Life expectancy<br>Diseases, morbidity, and mortality |
| Abrupt climate change | Possible significant impacts from rapid and extreme sea-level rise and temperature change | Access to water and sanitation<br>Durability of dwellings<br>Labour and employment |
| Population movements | Movements from stressed rural habitats | Security of ownership<br>Access to adequate living space |
| Biological changes | Extended vector habitats | Access to water and sanitation<br>Disease, morbidity, and mortality |

transportation system, resulting in cascade failures, such as problems with access to hospitals, energy supply, water availability, and rescue arrival. Emergency systems also rely on a functional transport system.

The increase of extreme events, such as heatwaves, intense rainfall, and sudden snowstorms, will directly impact transportation availability. Table 1.2 shows climate hazards that will likely impact the urban transportation system.

Low-carbon transport strategies involve a high-capacity mass transport system, more efficient multimodal networks, better vehicle design and cleaner propulsion technologies. This can also have other benefits, such as improving public health, better air quality, and redundancy of routes. Risk reduction solutions are also necessary, based on local risk assessment, supportive policy, and public–private participation.

As already mentioned, climate change directly impacts the so-called '***urban health***', including its buildings. Extreme heat, storms, floods, and landslides are among the originators of sequential impacts that can reach cities.

Besides infrastructure and geographic location, other factors that aggravate urban health vulnerabilities are the proportions of children, elderly, sick, and poor populations, which can vary drastically from city to city. Existing health conditions and diseases can also be exacerbated by climate change.

**Table 1.2.** Climate hazards and urban transportation impacts. Source: Adapted from Mehrotra and Zusman (2018).

| Climate hazard | Likely impacts |
|---|---|
| Intense rainfall and sudden snowstorms | Damage and incapacitation of transit networks due to the flooding of tracks, roadways, tunnels, and maintenance facilities<br>Blockage of roadways and drainage systems<br>Landslides because of precipitation |
| Heatwaves | Buckling of railroad tracks<br>Damage to paved surfaces<br>Passenger and worker health and safety |
| Sea-level rise and land subsidence | Flooding of roadways, tracks, maintenance facilities, and tunnels<br>Corrosion resulting from exposure to seawater |
| Typhoons, cyclones, and hurricanes | Coastal inundation from storm surge<br>Rainfall of sudden or intense nature<br>Damage from debris and high winds<br>Demand for evacuation services |

The growth of the urban population puts pressure on urban biodiversity and the ecosystem. ***Urban ecosystems and biodiversity*** are extremely sensitive to climate change effects, and some changes are already visible. Urbanisation and climate change are already increasing the vulnerability of urban species, habitats, and critical ecosystem services.

Urban ecosystems have a critical role in climate change adaptation and mitigation. They provide a natural environment between built spaces, with cost-effective nature-based solutions that can improve quality of life, human health, and social wellbeing. Urban ecosystems also promote sustainable urban development, a green economy, and social equity.

Temperatures rise can affect species' physiological processes, such as photosynthesis, respiration, growth, flowering, and plant development by altering growth and reproduction rates, either positively or negatively. Droughts can increase evapotranspiration, which can lead to reduced water availability and groundwater resources, with impacts on salinisation and water stress.

The shift of species in response to climate change has already been documented. For example, species can move north in search of cooler temperatures. Species with sensitive characteristics can suffer from local extinction. An urban health issue will be caused by species with unique attributes that benefit from climate change, such as the mosquito *Aedes aegypti*, a vector of various diseases.

Flood hazards can have impacts such as sediment movement, soil processes, and the distribution of pathogens that were once concentrated. In coastal cities, sea-level rise and floods can increase salinisation and reduce groundwater recharge, which affect habitat quality and biodiversity (IPCC 2014).

Modifications of species for any reason can also have cascading effects on the ecosystem, on other species, on soil fauna, or any other component of biodiversity; these can also have impacts on human activities and infrastructure.

The suppression of ecosystems is considered to be one of the most significant factors in reducing the resilience of cities. This is because they are more vulnerable to future problems that will be accentuated by global warming. Among the most important ecosystem services that can mitigate global warming are the provision of drinking water, the regulation of extreme events, the local climate, air and water quality, erosion, and carbon sequestration. Wrong decisions about ecosystems can involve huge losses of natural capital, for which the costs of restoration are much greater than the costs of prevention (PBMC 2016).

Water is an essential resource for life, humanity, and the economy. Good-quality water is fundamental for the increasing number of people living in cities and economic activities. Historically, many cities were located near water bodies; this applies to 13 of the 20 most populous cities and more than 70% of world trade.

Climate change imposes additional pressure, because temperature rises result in trends of increasing water consumption and hazards related to water. The excess or lack of precipitation, floods, pollution, droughts, and sea-level rise can not only damage essential infrastructure, but also reduce adequate water availability (Forman 2018).

***Water security*** is a central issue for cities. UN-Water (UNU-INWEH 2013) defines it as the ability of the population to have sustainable access to adequate water, to ensure protection against water-borne hazards, and to preserve ecosystems. This means conflict mediation, investments in infrastructure, and management between various stakeholders. For cities in low-income countries, this is particularly important, as many of them struggle to deliver basic services with a minimum quality and quantity of water. Moreover, as their populations grow, demand and competition will increase with climate change, leading to social stress and possible conflicts.

The impacts of climate change on water security can be categorised in two ways: water as a resource and water as a hazard. Increasing temperatures, changes in weather and precipitation, sea-level rise, storm surges, and changes in surface and groundwater availability are all threats that affect urban water security. Table 1.3 shows the risks that arise from climate change in terms of water as a resource and water as a hazard.

The adaptation to climate change must be built on a solid basis of good planning and governance, sources of investment, social equity in water access, and a resilient infrastructure made to support fast-changing conditions. Specifically, in low-income countries, the growth of the urban population imposes another challenge that requires improvements in the efficiency of the system and even the use of alternative sources of water.

Some alternatives for adapting water resources to urban infrastructure involve efficient water use, water storage, conservation, reuse, and rainwater use along with reviews and modifications of the surface and underground sources of water collection and transfer. It is also necessary to increase the water storage capacity and promote the recovery of hydrographic basins (PBMC 2016).

**Table 1.3.** Urban water security risks. Source: Adapted from Forman (2018).

| Parameter of water security | Risks |
|---|---|
| Water as a resource<br>Sustainable access to adequate quantities of acceptable quality water for supporting livelihoods, health, wellbeing, and economic development | Precipitation reduction, glacier retreat, land erosion, or sea-level rise leads to a decrease in water availability and/or worsening of water quality<br>Ecosystems are threatened by high stress caused to water resources by both climate factors and poor management |
| Water as a hazard<br>Ensuring protection against water-borne pollution and water-related disasters | A reduction in water flows could lead to an increased concentration of pollutants<br>An increase in precipitation intensity or storm surges could increase threat associated with floods |

The continuing trend of urbanisation, fast population growth, and climate change will impact the global ***energy infrastructure*** even more. The challenges for the energy sector involve three main issues: environmental impacts, energy access, and vulnerabilities to climate-related events.

The main resources of energy are still fossil fuels. Oil, gas, and coal support 84.9% of global energy use and are the biggest source of GHG emissions. The current trends of demographics and urbanisation will further increase the total amount of emissions. Although clean energy use is growing rapidly, it will still take a long time to change the game (Marcotullio *et al* 2018).

On the demand side, the warmer climate, the growth of urban populations, and global development put pressure on expanding energy use. On the supply side, extreme events and technological changes bring risks and opportunities for future development.

The problem is that energy emissions grow with the increase in the population and GDP of countries. The promotion of energy efficiency involves the complete decarbonisation of an economy's activities, reaching an almost 100% clean and renewable energy matrix. To achieve the goals of the Paris Agreement, countries such as Brazil should have net emissions equal to zero by 2050 and negative net emissions after that date.

Regarding vulnerabilities to climate-related events, three different approaches can be presented (Marcotullio *et al* 2018). The first is primary energy feedstocks related to risks such as reducing biomass crops to produce energy, which primarily affects developing countries, and oil and gas drilling operations, which can suffer from exposure to extreme events such as flooding and high winds.

The second vulnerability occurs in energy generation. It can be related to power plants located near coastal zones that can be damaged by sea-level rise and flooding. Another risk to the energy generation system is the increased amount of high peak demand with rising temperatures, which increases the risks of blackouts and brownouts. Also, droughts can cut off the supply of water, affect the cooling needs of plants, and provoke wildfires and high temperatures that damage infrastructure (Marcotullio *et al* 2018).

Renewable resource power generation can also be affected. Droughts and the decline of snow levels can affect hydroelectric facilities, an increase in cloudy days can weaken solar radiation by up to 20%, and lower wind speeds may diminish wind energy production.

The third vulnerability occurs in energy transmission and distribution networks. The hotter the temperature, the more conductivity declines, and power lines became more susceptible to failures. Extreme events, such as heavy rains, flooding, storms, high winds, and hurricanes also affect distribution systems.

Essential services depend on energy supply to work correctly. Therefore, reducing vulnerabilities is necessary to avoid cascading failures that threaten hospitals, transport systems, industrial production, and other critical human activities.

***Urban solid waste*** is intimately related to development, population growth, an increase in urbanisation, and climate change (Oteng-Ababio *et al* 2018). About 3% to 5% of global GHG emissions belong to improper waste management, most of which is methane produced in landfills. In developed countries' cities, for example, in North America and Europe, the waste sector is responsible for 2%–4% of total urban emissions. On the other hand, in less developed cities, like those in Africa and South America, waste sector emission corresponds to 4%–9% of total urban emission. Most of this difference belongs to the methane reduction infrastructure.

However, there are variations in the emission rates associated with waste in different cities, probably due to varying consumption patterns and generation of waste, differences in waste management, and differences in accounting mechanisms. The lack of a standardised urban structure that would allow inventories of GHG emissions can be conducted leads to variations that make it difficult to assess them (UN-Habitat 2011).

Climate-change-related impacts to waste sector infrastructure involve the three main processes of the system: collection, processing, and disposal. Table 1.4 shows climate change risks and their effects on each part of waste management.

Regarding the waste sector, there are many challenges in cities, but they all point in the same direction, which is to increase recycling, reuse, and reduce waste generation. Better landfills with energy recovery systems can also help to reduce local pollution and GHG emissions. For low- and mid-income countries, population growth, an increase in per-capita generation, and poor infrastructure poses great challenges. Improving solid waste management can also create opportunities for collection, recycling, and landfill building and maintenance.

## 1.4 Sustainable city responsibilities

The environmental standards of cities must shift as global change impacts take place. The increase in the intensity and frequency of extreme events will provoke disasters and climate emergencies that cities will need to respond to in the coming decades. Many aspects of these are related to vulnerabilities, such as culture, demographics, economy, governments, technology, the built environment, ecosystems, resource exploitation, and degradation. Cities that do not adequately plan their urbanisation face serious threats, because environmental change can remove natural barriers against storms, increase air and water pollution and the heat-island effect, and exacerbate the overall impacts of climate change.

The transition to low-carbon cities is critical to the sustainability movement. The relation between urbanisation and GHG emissions is complex. Cities are responsible for a great part of GHG emissions, most of the world's population, and its growth. However, they present opportunities to efficiently use resources in large populations, focus on sustainable development goals, reduce carbon emissions, and prepare for upcoming events.

It is fundamental that cities must adapt to global change. Climate change, migrations, and the growth of urbanisation are strong trends; their impacts are already materialising and demanding city action. Making cities and human settlements inclusive, safe, resilient, and sustainable is essential in order to respond to the global change and its impacts and to achieve the SDG 11 'Sustainable Cities and Communities'. Transformative climate change measures create the possibility of

**Table 1.4.** Climate change impacts to the solid waste management sector. Source: Adapted from Oteng-Ababio *et al* (2018).

| | Collection | Processing | Disposal |
|---|---|---|---|
| Temperature change | Increased odour, requiring more frequency<br>Overheating of collection vehicles | Overheating of sorting equipment | Altered decomposition rates<br>Increased maintenance and construction cost<br>Increased risk of fire |
| | Greater exposure of workers to flies, increasing disease | | |
| Precipitation change | Flooding of collection routes and landfill access<br>Increased stress from waterlogged waste | Increased need for enclosed or covered sorting facilities | Increased flooding in/around sites<br>Increased leachate<br>Risk of fire if too dry and hot |
| Sea-level rise | Narrowed collection routes<br>Potentially increased waste in a concentrated area | Damage to low-lying processing facilities<br>Increased need for sorting and recycling to minimise waste storage | Deterioration of impermeable lining<br>Water infiltration of pit leading to possible overflow of waste |
| Storm surge | Permanent inundation of infrastructures<br>Temporary flooding and diminished access to roads and facilities<br>Closure of facilities due to infrastructure damage | | |
| Extreme wind | Dispersal of waste from collection sites, collection vehicles, processing sites, and landfills<br>Reduced access to routes due to damage and debris | | |

generating more extensive moves toward sustainable cities. In this way, three fundamental characteristics emerge for adaptation and mitigation improvements: resilience, efficiency, and inclusivity.

The concept of efficiency is linked to economic theory and comprises the management of resources. It is a way of expressing the level at which a nation's known production capabilities are taken advantage of. The relationship between the effect, the effort, and the results achieved, in terms of a specific period or other best practice define the economic efficiency (Borza 2014).

With the advent of the COVID-19 pandemic, the ideas for the economic crisis recovery have evolved into concepts more appropriate to fight against climate change and its consequences. In this sense, the idea of a Green Recovery Plan has arisen, which consists of a transition strategy for a green economy which is compatible with the fight against inequality, with the generation of employment and income, and with sustainable economic growth. Thus, the concept has established itself by considering the need to reduce or eliminate the economic activities that strongly drive carbon emissions.

The concept of efficiency is also directly correlated to sustainability. To maintain and improve people's wellbeing, it is imperative to do more with fewer resources. The term 'eco-efficiency' complements the efficiency concept by adding the environmental resources and impacts to the equation. Eco-efficiency is defined as the efficiency with which ecological resources are used to attend to human necessities. Eco-efficiency can be expressed as the ratio between the amount of social service provided and the environmental burden generated. For urban areas, eco-efficiency is described as the ratio between citizen wellbeing and the ecological burden caused by the city, so efficient urbanism is achieved when the social services increase more than the environmental or ecological burden.

The city's answer to the increase in the frequency and severity of extreme events must include a new decision framework and disaster risk reduction measures that build resilience against these impacts. Resilience is a concept first related to ecosystems and how they are applied to urban space.

Resilience is a crucial aspect for mitigating the impacts on the infrastructure due to extreme events. Therefore, every infrastructure system must assess its vulnerabilities and their interactions to improve the city's resilience. Resilience can be measured as a combination of three factors: the quantity of disruption a system can absorb and still remain in the same state, the capacity of the system for self-organisation, and the ability to build and increase capacity for learning and adaptation (UN-Habitat 2017). Other critical aspects for cities are the ability for the municipality to develop long-term planning strategies, management processes, and urban system efficiency.

For low- and mid-income countries, inclusivity plays a special role. The rapid urbanisation process has led to enormous migration from rural areas to cities, which added to the liberalisation and privatisation of public spaces and infrastructure, creating pressure on public services and resulting in poor and informal settlements that decreased the overall quality of life. Additionally, bad infrastructure, poor-quality housing, and the lack of education, jobs, and public services contribute to increased socio-spatial and financial inequality, further expanding the gap between rich and poor people. This creates the poverty trap that puts pressure on infrastructure systems, stressing earning potential, political participation, and population wellbeing.

In poor cities and cities with great inequality, global change can have a magnified impact in comparison with the impact to rich and organised cities, making it difficult to take some types of mitigating measure. For example, the lack of ownership in informal settlements can limit any action of evacuation; since people do not know if they will ever return home, they can hesitate to take this decision.

Impacts on low-income populations can have additional cascading effects. For example, the low quality of dwellings can increase disease and mortality due to extreme weather events or vector-borne disease. Otherwise, the same can be said for the upside effects of mitigation measures. For example, an improvement in public transportation can reduce activity costs and the time spent in traffic and improve population wellbeing.

Mitigation and adaptation efforts must prioritise measures that encompass multiple key aspects. Urban planning decisions have long-term consequences and a high potential to reduce GHG emissions and respond to climate hazards. Sustainable urban planning, in conjunction with regional climate change assessment, can improve urban environments and address resilience, efficiency, and inclusiveness.

According to the IPCC (2014), there are some necessary mitigation efforts. In the area of energy supply, the use of low-carbon technologies including renewables, nuclear sources, and carbon capture and storage can reduce GHG emissions. Energy demand can be reduced by up to 20% in the short term and 50% by 2050 by increasing the efficiency of buildings and achieving changes in residents' awareness and behaviour. For transport, emissions can be reduced by moving to low-carbon transport systems, increasing the efficiency of vehicles and engines, and reducing the carbon intensity of fuels by the use of engines powered by electricity or hydrogen. As for buildings, the renovation of existing constructions can reduce heating and cooling energy needs. New buildings in rapidly growing regions represent an opportunity to mitigate the effects of global warming; in such buildings, GHG emissions can be virtually eliminated. In cities undergoing rapid development, the trajectories of urban development and infrastructure must be shaped. For mature cities, the options are in urban regeneration and rehabilitation.

## 1.5 Conclusions

This chapter has presented how global change can take place and will affect the urban environment, infrastructure, and services. Even if limited to the Paris Agreement level, 1.5 °C above pre-industrial levels, global warming can cause enough damage to change ecosystems and harm populations profoundly. Although it is still somewhat uncertain whether that level will be achieved, studies of a possible 2.0 °C global warming show that most of the projected impacts would significantly increase. Therefore, immediate efforts must be made to avoid an increase in temperature above the proposed levels.

Global warming effects are not uniform and not precisely known. Although the global mean temperature will rise, this increase will be heterogeneous, causing different impacts. Moreover, the combination of effects has the potential to multiply risks. For example, the rise in ocean temperature can raise sea levels, melting glaciers and ice sheets, which leads to even more sea-level rise. In addition, the increase in temperatures and the lack of precipitation can lead to droughts, and, at the same time, increase water consumption in cities, which can aggravate the situation.

The increase in the frequency and severity of extreme events will be a challenge for humanity, especially in low-income cities. Cities must develop strategic plans and prepare for the impacts by improving resilience, efficiency, and inclusivity. Resilience must be built by taking account of the whole city, avoiding cascade failures and resource wastage. Urban planning that fails to adjust zoning, building codes, and standards for the climate-change future will limit the possibilities for adapting the city's infrastructure and put the lives of its citizens at risk.

Low- and mid-income cities face additional challenges due to increasing urbanisation, social and economic inequality, migration, and informal settlements. In addition, many of these cities are in areas that are more vulnerable to climate change impacts, making adaptation and mitigation efforts even more crucial. Climate action in such places must also consider inclusiveness and efficiency as essential aspects for development, improving people's lives, and offering a path towards sustainability.

## Bibliography

Borza M 2014 The connection between efficiency and sustainability—a theoretical approach *Procedia Economics and Finance* **15** 1355–63

de Alvarez C E and Bragança L 2018 Medidas para o Enfrentamento dos Impactos das Mudanças Climáticas no Ambiente Construído *SUSTENTABILIDADE URBANA: 14ª Jornada URBENERE e 2ª Jornada CIRES (Vila Velha, Brazil), Proc.*, V1 3–12

Forman R T T 2014 Urban water systems *Urban Ecology: Science of Cities* (Cambridge, UK: Cambridge University Press) ch 6 149–74

Gray V 2007 Climate Change 2007: The Physical Science Basis Summary for Policymakers *Energy and Environment* **18** 433–40

Field C B, Barros V, Stocker T F and Dahe Q (ed) 2012 Summary for Policymakers *Managing the Risks of Extreme Events and Disasters to Advance Climate Change Adaptation: Special Report of the Intergovernmental Panel on Climate Change* (Cambridge, UK: Cambridge University Press) pp 3–22

IPCC 2018 *Global Warming of 1.5 °C. An IPCC Special Report on the Impacts of Global Warming of 1.5 °C above Pre-Industrial Levels and Related Global Greenhouse Gas Emission Pathways, in the Context of Strengthening the Global Response to the Threat of Climate Change*, ed T Waterfield Masson-Delmotte, V P Zhai, H-O Pörtner, D Roberts, J Skea, P R Shukla, A Pirani, W Moufouma-Okia, C Péan, R Pidcock, S Connors, J B R Matthews, Y Chen, X Zhou, M I Gomis, E Lonnoy, T Maycock and M Tignor (Geneva: IPCC)

Masson-Delmotte V P *et al* (ed) 2021 *Climate Change 2021: The Physical Science Basis. Contribution of Working Group I to the Sixth Assessment Report of the Intergovernmental Panel on Climate Change* (Cambridge, UK: Cambridge University Press)

Jean-Baptiste N, Olivotto V and Porio E *et al* 2018 Housing and informal settlements *Climate Change and Cities: Second Assessment Report of the Urban Climate Change Research Network* (Cambridge, UK: Cambridge University Press) ch 11 399–431

Marcotullio P J, Sarzynski A and Sperling J *et al* 2018 Energy transformation in cities *Climate Change and Cities: Second Assessment Report of the Urban Climate Change Research Network* (Cambridge, UK: Cambridge University Press) ch 12 443–90

Mehrotra S and Zusman E 2006 Urban transportation *A Companion to Urban Economics* (Hoboken, New Jersey: John Wiley & Sons, Ltd) 241–4 Part IV

Oteng-Ababio M, Annepu R, 'Thanos' Bourtsalas A C, Intharathirat R and Charoenkit S 2018 Urban solid waste management *Climate Change and Cities: Second Assessment Report of the Urban Climate Change Research Network* (New York: Cambridge University Press) ch 6 175–80 https://doi.org/10.1177/0734242X8800600126

Ribeiro S K and Santos A S 2016 *Mudanças Climáticas e Cidades. Relatório Especial do Painel Brasileiro de Mudanças Climáticas* Painel Brasileiro de Mudanças Climáticas (in Portuguese)

UN 2020 *Goal 13: Take urgent action to combat climate change and its impacts.* https://un.org/sustainabledevelopment/climate-change/

UN-Climate Change 2019 *The Paris Agreement* https://unfccc.int/process-and-meetings/the-paris-agreement/the-paris-agreement

UN-Habitat 2011 *Global Report on Human Settlements 2011: Cities and Climate Change* (UN-HABITAT)

UN-HABITAT 2016 *World cities report 2016—urbanisation and development: emerging futures*

UN-Habitat 2017 *Trends in Urban Resilience 2017* (UN-HABITAT)

UN-News 2020 *Flagship UN study shows accelerating climate change on land, sea and in the atmosphere.* https://news.un.org/en/story/2020/03/1059061

UNU-INWEH 2013 *Water security and the global water agenda. A UN-Water analytical brief* 53 (Hamilton, Ontario: United Nations University Institute for Water, Environment & Health (UNU-INWEH)) https://collections.unu.edu/eserv/UNU:2651/Water-Security-and-the-Global-Water-Agenda.pdf

# Chapter 2

# Project logic for the reorganization of cities in transition: a Hispanic American application

**Fermín Rodríguez and Alvaro Valverde**

The application of a project logic for the reorganization of cities in transition as a dynamic and inclusive process encourages continuous improvement in the formulation and evaluation of environmental governance policies and practices that tend to generate better options in the achievement of shared management programmatic goals. These goals are considered to be multicriteria interventions, which generate a participatory learning cycle based on reactive or proactive actions practiced by the agents involved in urban management. Such a learning cycle links all decision processes to the approach of 'learning by doing', in which different knowledge systems are synthesized, projecting a flexibility aimed at collaboration and cooperation between management at local, regional, and national levels, adopting the city as a territorial planning unit in constant transition, whose substantive element is to specify a model city.

## 2.1 Introduction

The city has faced problems with solutions obtained according to temporal patterns which claim to achieve a model city, a constant that accompanies man as a territorial being, so that the result in each stage is judged according to the objectives of the civilization that raises it. The claim that building urban artifacts is an expression of citizen values is one of them, since it has classical roots, it is identified with a humanist claim, and it does not follow a path of linear progress, but there may be regressions depending on the mistake, with which each city solves the problems raised.

At present, there is also no single approach to the concept of the model city; however, it is the attribution of a series of values for stable human wellbeing that guides its management. It is not the object of this article to look for a project model with which to focus on the transformation of cities in transition, but to investigate a

doi:10.1088/978-0-7503-3971-1ch2  

project logic with which to reform them, based on our understanding of them as complex and dynamic citizen systems, for whose operation a multicriteria analysis seems suitable.

We consider that a human settlement of a certain size does not make a city; no matter how organized it may be, it does not qualify as a civitas, as a space of good government, in which certain values are expressed: freedom, equality, security, and autonomy in a physical form, because with similar populations, one settlement can be a concentration camp and another, a territory; this is an operational concept that helps us to guide the action of agents towards the creation of a physical space as an expression and product of those values. Neither is the city a productive device that pretends to be a large factory, in which the physical layout is almost a matter of small-scale ergonomics, aimed at promoting production and optimizing the productivity of the workers housed there.

The city is the expression of a pretense of good coexistence, a physical state of civilization like an enlightened stadium, a territory of physicosocial construction which, due to its degree of artificiality, needs to incorporate, for its progressive stability, a delicate game of balance between agents and citizens.

Therefore, the ideal parameter with which to guide the construction of the city would be one that would guarantee friendship between citizens, since 'friendship is the greatest asset in cities', which we will translate here by acting on the public spaces that allow harmonious interaction with each other and with the environment, without violence, and in an equitable order. To do this, we use a strategic framework to identify the general and specific attributes of a city, which implies understanding what makes it viable, how it is transformed, and how it achieves a degree of resilience in adaptation or mitigation contexts.

The systemic need is consequently fueled by the current global challenges that large institutions and organizations spread in the consciousness of the masses, who respond with urban governance decisions based on the paradigms of: sustainability, development, and resilience, or their default: economic sustainability, growth, and adaptation/mitigation.

This chapter defines a project logic as its own approach to urban spatial planning, in which the integration of multiple criteria, the search for priorities in the pertinent actions and the prospect become significant concepts. The first is the property of transversality, which allows us to incorporate analyses from different disciplinary fields into a theoretical scheme in order to integrate them into an interdisciplinary theoretical corpus, built for intervention in the territorial system by specialist physicians considered to be more like knowledge smugglers than infallible technocrats; the second means that all the systemic components of the structure have the same weight of relevance. Finally, prospective foresight orients future processes towards the anticipation of consistent states.

The project logic starts from the establishment of the current state of the territory, considered as a functional and symbolic system, embedded in a succession of scales. For its diagnosis, one needs to integrate two approaches into the analysis: the functional and the hermeneutical; and for its ordering, encouraging a set of actions that will drive it to achieve a new state from both sequential and simultaneous applications.

Taking into account that the evolution of the system is carried out in successive stages, each one of which contains traces of the previous one, urban control strategies will be directed towards teaching–learning objectives, both for those involved—as a way of transmitting a territorial culture—as well as for the specialist physicians and territorial evaluators, who will act in the accompanying stages of the integrated urban reform process: (a) helping to define its purpose, (b) establishing programmatic objectives that become goals, (c) incorporating processes, methodological procedures, and resources, and (d) determining an implementation and evaluation format appropriate to the geographic and biocultural context of the urban space, in an interesting contrast that will be studied in both Bolivian and Spanish cities. We consider that in both, there are metropolitan, intermediate, and emerging conditions marked by a series of similarities and differences; that is to say, a mix of expertise will emerge as a result of combining the domain of the territorial analyst's methodology with that of the diffuser of the consequences of the agents' actions that can be anticipated, handling themselves on both levels as functionalists and as hermeneutics; and such management must be done from a proper deontological perspective.

In order to maintain a necessary coherence with other scales and with an awareness of the advantages obtained by being able to compare case studies within the proposal, the World Report on Sustainable Development for 2019 and 2020 (carried out by the Bertelsmann Stiftung Consortium) will be used as a documentary reference, in which the most approximate assessments established regarding the Sustainable Development Goals (SDG) in each country are made, as well as the institutional experiences that have marked a significant trend for the writing of this chapter. The authors of this chapter are linked to the CeCodet of the University of Oviedo and the Bolivian Catholic University and, as members of these institutions, have participated in applied research projects, both in Bolivia, where the application of a proposal for cities in the highlands (La Paz), the Chaco (Tarija), and the valley (Santa Cruz) were studied; and in Spain, where the reference case is that of the city of Mieres, the focus of activity for the EDUSI—Mieres project, which seeks to reorient the restructuring of an industrial and coal city, based on an integrated urban regeneration approach. This creates an interesting framework for action, given the well-marked differences in geographic, economic, and cultural terms between the two countries, although the interventions have a similar orientation and methodology.

## 2.2 Discussion of urban transition

Development is a constant of territorial systems. This word is generalized and mystified. For this reason, it would be more appropriate to understand it as a restructuring or a period of accelerated changes with far-reaching consequences that lead to a new phase or cycle, since, unlike biological systems, territories do not die, but precisely because of their inherent symbolism and their complex nature, they do not have meta-states such as cybernetic systems. Indeed, social engineering has found important allies to help such definitions of goals in the new information technologies, which, if programmed without participation and free, voluntary, and informed acceptance, would disqualify the territory concept by eliminating its attributes: freedom, security, and autonomy for individuals, who thus acquire the condition of citizens, protecting the dignity of each person.

There are many versions and nuances in defining a single approach to the concept of 'a model city', so it is much more successful—for now and until its adoption—to conceive a project logic that triggers a series of multicriteria interventions within the structure and function of cities in transition. Given that each city responds to its own geographical, climatic, biocultural, sociopolitical, and economic–productive characteristics, it is possible that a strategic framework of inquiry and the determination of priorities will enable the identification, adoption, and redefinition of a series of reactive and proactive actions for the constitution of an approximate urban model.

In this urban transition path, cities promote dynamic and inclusive processes that the project logic proposal incorporates to encourage urban regeneration, in an attempt that also promotes the formulation and evaluation of urban environmental governance policies and practices aimed at generating greater and better programmatic goals of shared management. These goals are considered to be multicriteria interventions (policies, strategies, and actions) that create a participatory learning cycle based on responses and attentions practiced by the agents involved in the short, medium, and long term. These interventions are linked in all urban decision processes, under the collective approach of 'learning by doing', from which different knowledge systems are synthesized that project a flexibility oriented towards collaboration and cooperation in management between local, regional, and national levels. The national level, adopting the city in transition, is a substantial element for the logic of the territorial intervention project.

The foregoing implies that the project logic should concentrate efforts on recommending: (a) responsible and transparent institutions, which require new capacities, forms, and practices of governance, (b) the provision of greater communication between the actors involved regarding the problems resulting from the environment—production relationship, (c) the study of shared visions and binding strategies to reach consensus among stakeholders, accelerating the promotion and implementation of processes with the direct participation of community members, especially youth and women, in the search for solutions; but above all, (d) the promotion of a system of strategies that investigate, prioritize, and project the trend of operational measures that respond to the problems evidenced in cities, while articulating effective urban governance.

This last condition is based on the need to provide a project logic on which an evolved model of governance is constituted, supported by the dichotomous trajectories of choice for each city (in the case of Spain and Bolivia), following the direction of two currents that are today well established: on the one hand, sustainability, development, or resilience, and on the other, economic sustainability, growth, and adaptation/mitigation. Table 2.1 shows an approximation of the proposed categories and highlights the differentiation between the two possible transitions.

In contrast to the realities and needs identified in both Bolivian and Spanish cities, the concern and motivation of the proposal *project logic for urban territorial intervention*, refers to rethinking cities in transition at the systemic, comprehensive, and multidimensional scales, considering actors, sectors, and disciplines, which makes possible the exercise

**Table 2.1.** Approaches to dealing with urban transition (Valverde *et al* 2020).

| Transition 1 | Transition 2 |
|---|---|
| *Long term and reduced space* | *Short term and large space* |
| *Sustainability.* The possibility of maintaining biocultural, economic–productive, and sociopolitical processes during generational periods, thereby obtaining better resources and results with an equipotential distribution for all populations, especially those that are fully localized as vulnerable. | *Economic sustainability.* The ability to achieve social and economic wellbeing by estimating its immediate effects on the natural and cultural environment of the environment, obtaining from it greater resources and results that are valued according to their availability, access, and use to enhance the survivability of populations. |
| *Development.* A consensus-mobilizing project due to the sociopolitical construction, the economic–productive transformation, and the environmental balance that redefines the enhancement of the potentialities and limitations of the environments and their resources, to take them in the long term to a qualitatively better state. | *Growth.* A condition that increases economic activity with social satisfaction through the aggregation of available material, financial, technological, and energy resources, to bring it in the short term to a quantitatively higher state. |
| *Resilience.* A capacity attributed to the values and qualities of an individual or a community, which overcomes any adverse condition and recovers from it with integrity, overcoming challenges, and emerging strengthened from negative events, to contribute to the transformation of its biocultural, sociopolitical, and economic environment—productive in the medium and long term. | *Adaptation/mitigation.* Individual or collective intervention that creates, modifies, adopts, attenuates, or increases the dynamic recovery of socio-natural systems, due to changes, uncertainties, complexities, and conflicts generated by threats and vulnerabilities in the immediate environments, mobilizing human, material, and financial resources to implement effective actions. |

of effective urban governance, according to the direction that cities are going to assume in the management processes. Indeed, figure 2.1 shows a scheme of the elements considered for the application of the project logic for urban reorganization.

Therefore, the backbone content from which complex but dynamic knowledge is built, which sustains the proposal on a strategic basis with a prospective approach, establishes a triangulation of theories, experiences, and projections that generate a space for self-critical and reciprocal inquiry in the study of urban cities, which circulate through cycles, phases, domains, and organized or regulated axes, so that the knowledge of the systemic nature of the processes acquires greater depth with the approach of urban responses, from seeing the imaginary built to anticipating the future, foreseeing the present and overcoming the past.

Both the inquiry and the determination of priorities in urban strategic decisions and actions must create a framework to integrate the structural axes on which to specify a project logic for cities in transition. In such a framework, the multicriteria

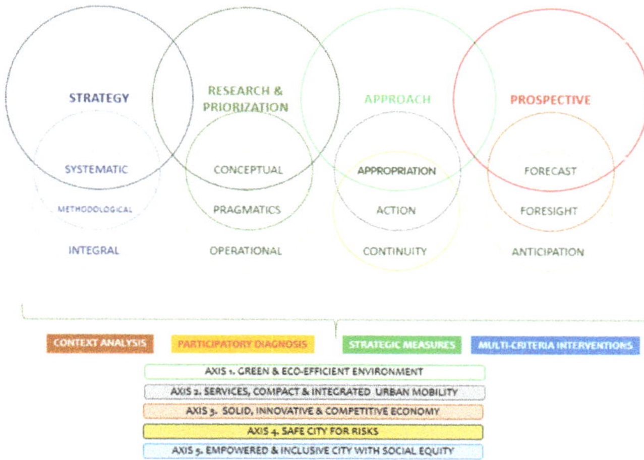


**Figure 2.1.** Scheme of the elements of the project logic for urban reorganization (Valverde *et al* 2020).

dimensions (actors, scales, sectors, and disciplines) adopt a significant role, as they will allow us to face the complexities and uncertainties of the urban system, helping to define the criteria that underpin the different phases of the evolutionary cycle of a city (geopolitical, technological, adaptive, regenerative, and social) and which are decisive when setting short and long-term objectives. Such objectives depend on the choice of actions as well as the allocation of resources necessary to achieve them, so that each strategy supposes a process that is essentially purely intuitive, such that it is territorial planning understood as an enlightened function and not as social engineering.

## 2.3 Methodological development of the proposal

### 2.3.1 Purpose

The proposed project logic for urban intervention in cities in transition particularly underlines the possibility of carrying out a series of orientations and concrete actions within a program that lead towards the creation, strengthening, and updating of cognitive, attitudinal, and procedural competences in the actors involved in establishing an efficient and effective territorial planning model, and will consider promoting a greater diversity of spaces for relevant discussions that take place in a context marked by academic, technical, and political interrelationships, synthesizing the best experiences that contribute to the realization of the proposal and in which the significant role of the territorial evaluator must be guaranteed by their autonomy, optional expertise, and their understanding of the system of symbolic and functional relationships, taking as a reference the inquiry and presentation of the consequences that can be expected from the integrated result of the participants' actions.

### 2.3.2 Objectives

When considering the strategic prospect as the basis of the approach used for the inquiry and the prioritization of urban interventions, the relationship of the domains

**Table 2.2.** Stages, domains, and goals of the proposal (Valverde *et al* 2020).

| Stages | Domains | Goals |
|---|---|---|
| Context analysis | Mapping of actors and objectives | Synthesize the degrees of convergence and divergence between actors involved in the study, as well as mobilization actions to respond to prioritized and non-prioritized objectives. |
| Participatory diagnosis | Reflections on the commitment to cities in transition | Synthesize the results of consistency and trend of the diagnosis developed, considering the directionality of each city (economic sustainability vs sustainability, growth vs development, resilience vs adaptation/mitigation). |
| Strategic measures | Alternatives and chosen inferences | Establish adequate structural and functional contents for the proposals, which provide an emergent and programmatic response to environmental, urban, and institutional needs. |
| Multicriteria interventions | Scopes according to scenarios | Project the scope of continuity of the prioritized measures, after evaluation of scenarios and conditions regarding available resources (economic, human, and material). |

that are of interest for promotion are those that will be applied for each city in the Bolivian and Spanish geographical areas, with their respective differences; table 2.2 shows certain common stages, domains, and goals.

### 2.3.3 Processes, procedures, and resources

Urban planning is, by nature, a complex process, the result of which is inherently subjective and uncertain, since, like its object, the city, it is vulnerable. For this reason, the success of urban planning is measured by the combination of its technical soundness and its execution possibilities. It is a matter of balance, of seeking harmony in decisions. The study consequently intends to address, using multicriteria planning, the determination of inferences (characteristics, qualities, and conditions, among other minimum considerations), which should incorporate the most appropriate concept for a city in transition from its sustainable economic or ecological choices, oriented towards growth or development, and its link to actions that specifically support adaptation/mitigation or a resilient/adaptive character.

Since traditional planning is essentially retrospective, because the objectives to be pursued are determined by what happens in the present, in the prospective mode the opposite occurs, since the desired future is established as openly as possible, to enable creativity and freedom from the constraints of experience and current circumstances. This implies rethinking territorial planning from a situation full of uncertainties, in which the effects of economic and environmental crises are an increasingly frequent reality.

Several cities have had the experience of opting for the use of foresight as a discipline with which to establish the logic of an urban territorial project

(Inter-American Development Bank's Emerging and Sustainable Cities Program) and have achieved greater development and better competitiveness in their environment by mobilizing all of the local actors as part of a co-managed planning model to prioritize a convergence of intent that can accelerate strategic actions accompanied by territorial policies. Following the above, the proposed urban territorial project logic considers the following elements sequentially and/or simultaneously:

- Identify the categories of criteria (scales, actors, sectors, disciplines, and approaches) of greatest importance in the territory, within a defined horizon of at least 10 years.
- Specify the key domains (indicators, attributes, variables, and measures) that will define the future of the territory.
- Establish interventions and strategic policies for the occurrence of the best urban scenario in transition.
- Build a conceptual and operational framework of a model city.

Therefore, the objective of the proposal is the design of a concept based on the inference of probable (deterministic) and alternative (betting) scenarios. The foregoing allows us to distinguish, with greater or lesser consensus, the following conditions and to specify the project logic proposal:

- Long-term approach: provides the stakeholders with information on possible scenarios for the evolution of sociopolitical, economic–productive and biocultural issues that are not always considered in current territorial planning practices, emphasizing long-term objectives that favor a collaborative attitude, reducing conflicts in daily operations.
- Multicriteria approach (scales, actors, sectors, disciplines, and approaches): opens the participation of actors with different points of view, competencies, and cultural bases, thus facilitating reciprocal enrichment and adding value to the process of identifying objectives and priorities.
- Greater attention both to the process and the results: thus, the vision-building exercise becomes a collective learning process that is nurtured by the union of different institutional, economic, and social competencies and actors. A planning process can become an occasion that induces a change in the behavior of the actors considering the outcomes to be reached, instead of only identifying priorities and objectives. Scrupulous compliance with the method is emphasized, since the result of the project cannot be known in advance; the result is the application of its logic, understood as a process of expertise and optional values that approximate the objective to a certain extent.
- Convergence towards one or more objectives and/or possible scenarios: the prospect is an iterative process that begins by defining an objective and tries to shape actions and use resources in the present to work towards its realization. It satisfies the need for public and private actors to share a vision of possible futures (thereby avoiding the formation of simple associations focused on specific interests) and becomes an opportunity to learn and increase collective knowledge and local intellectual capital, forming active opinion platforms.

### 2.3.4 Project implementation

#### *2.3.4.1 Context analysis*

Cities are the product of a series of changing biocultural, sociopolitical, and economic–productive conditions that are complex in their organization because they can host economies that support livelihoods and must connect the human side with the natural dimension. Due to their exponential growth, they are constantly modifying and reinventing themselves, eager to unfold their urban potential; for this reason, a 'city in transition' is a place where the structures and functions of urban environmental management are conceived and reconfigured; this process, in which the open prevails over the rigid and flexibility prevails over rigour, is even more important than the construction of the architectural body itself.

The recognition of these challenges in 'cities in transition' plays a new role in the contemporary urban world directed towards the construction of new and updated intervention paradigms, of new operational platforms, and of an activism for the 'transition condition' in the recognition of a new conceptual, strategic, and operational model that the population recognizes. This new theoretical–practical construction is the material of interest to be specified by the proposal; in this context, the primary challenge for cities in transition is to create appropriate negotiation and dialogue platforms for the analysis and formulation of responses and solutions, in time to postulate a series of prospective methodological implementation mechanisms which ensure an eloquent materialization of ideas, programs, plans, projects, and strategies for cities.

The characterization of the city, then, is not defined by the average that the indicators show in each of the proposed topics, but by the individual evaluation of each structural criterion of the city model, whose process depends on the greater degree of intersubjectivity than all the actors. Those involved weigh each city, following qualitative and quantitative hypothetical evaluations of the critical issues between internal areas of the territory, which, in some cases, will show similarities and differences for later treatment.

Therefore, it is proposed that in the future, territorial analyses be carried out that define large urban groups, starting from the center in some cases, with a characterization based on continuity and building density, and trying to identify the limits of the gradation of the urban landscape to the periurban and even rural areas. This transition will be complex, from cities where there is no gradation, i.e. compact cities with defined urban limits (with obvious separation), to complex cities where gradation is complicated by the appearance of diffuse urban growth.

Returning to the context analysis, this depends on the combination of elements that can be identified under multiple scenarios; however, the study focuses on three possibilities: trend (spatial projection), optimal (by analogies with other territories to which it is desired to aspire or by identifying goals for urban standards), and an intermediate one (based on the consensus of the majority of political and citizen wills) that is called 'viable' or 'consensus'.

It is precisely the middle, based on degrees of consensus, where performance, interest, and viability can be evidenced as the most important quotient with which to start the context analysis. For this point, we start from an institutional mapping of key actors and objectives which considers the competencies and attributions that those involved in urban issues have. Indeed, with the application of strategic territorial prospective software called MACTOR (whose acronym means: Matrix of Alliances, Conflicts, Tactics, Objectives, and Recommendations), we seek to assess the relationships of force and study the degrees of influence and dependence between actors in relation to the discussion topics to be raised, to later describe the convergences and divergences with respect to a certain number of positions and perceptions associated with the structure of axes, components, indicators, and objectives, which are initially proposed.

The structure of the proposal in 'urban territorial project logic', considers five strategic axes, which are, in turn, composed of 16 specific components, due to each axis, which were considered based on the experience of results in Bolivian and Spanish cities; therefore, its definition establishes the following foundations:

- Green and eco-efficient environment: the management and conservation of the natural environment of the city is the relevant basis for guiding an eco-efficient urban proposal. The environmental axis proposes developing efficient and participatory management mechanisms, generating environmental awareness and citizen involvement, and taking advantage of the benefits of nature to solve urban problems.
- Services and compact and integrated urban mobility: the trend of dispersed and uncontrolled growth of the urban area is a problem that makes cities and places segregated and inefficient for distribution and the supply of transport. This axis proposes controlling the expansion of the urban area, maximizing the positive impact of the construction and implementation of mass transport, and establishing an integrated and efficient mobility system that encourages active mobility.
- Solid, innovative, and competitive economy: the construction of economic resilience starts from the strengthening of the productive sectors and from the diversification of business lines in an environmentally responsible manner. This axis proposes the generation of an economic environment conducive to strengthening labor supply and demand, with a focus on young people and women; promotes a diverse, sustainable, and innovative economy; and promotes the food economy as an axis of development.
- A safe city prepared for risk: multiple threats and the high level of physical and socioeconomic vulnerability of the city mean that risk is distributed in a heterogeneous way in the territory. This axis proposes avoiding the creation of new risk, mitigating existing risk, and preparing the city to face latent natural and anthropic threats.
- Empowered and inclusive citizenship with social equity: this responds to the need to consolidate participatory processes as a vector of democratic exercise to validate the work of the public administration and facilitate processes of co-responsibility between citizens and the city. It aims to strengthen institutional and community capacities to strengthen participatory processes and provide clear and effective mechanisms that are reflected in the public space.

Consequently, the context analysis is realized by conceiving, from the beginning, each of the proposed axes, components, and indicators (60) that are linked to these foundations, and, according to the objectives set—oriented to management processes or institutional mission—we evaluate the favorable, neutral, or opposing perceptions of the actors involved, who clearly have competencies, interests, and positions according to their own involvement in the issues raised.

As part of its operation, MACTOR highlights the most intersubjective way (close to objectivity, admitting subjectivity of judgment) of facilitating the decision of the actors regarding their true actions regarding the issues raised, making it clear that they even acquire a certain awareness at the time of assessing a suitable weighting for their evaluation. In other words, the so-called identity card of each actor is established: their purposes, objectives, projects in development and maturing (preferences), their motivations, obligations, and internal means of action (coherence), and their past behavior strategic (attitude). The means of action that each actor has, with which they carry out their projects, are also compared to those of the others, understood as the achievement of their strategic objectives.

In this way, a qualitative and quantitative database is prepared in a way that is broadly necessary to continue with the processes of the next stage (participatory diagnosis), and which additionally considers another type of complementary convenient information. The interpretive and analytical reading does not follow a properly objective rigour, since the tool used aims to show a first approach to the approximate reality regarding the degrees of convergence and divergence, which is nothing more than the average of the net relationships between actors and objectives that show more or less attention, interest, and compatibility. For this reason, it is important to turn to other sources, mostly institutional and bibliographic, to know the actions of each actor in the face of the issues raised in the context analysis (mission, vision, academic and institutional production, successful projects, and lessons learned, among others). With this, possible scenarios can be analyzed when diagnosing the reason for the resulting alliances and conflicts and promoting measures that will seek change in the future.

Table 2.3 shows how the context analysis is carried out using different phases: (a) institutional registration, coding, and description of the mission/vision of the actors in each city. For this, an analysis of the capacity for action, the interest, and mobilization of each actor, their goals and objectives, as well as their strengths and weaknesses is carried out; and (b) the definition of strategic objectives: a registry of components, indicators, and the set of actors. For this, a study structure is established, conceived on a minimum and coherent basis of optimal criteria, which mark a critical path of initial action.

Finally, the application of MACTOR implies the presence of a certain number of limitations, mainly in the area of obtaining the necessary information. The reluctance of some actors to reveal their real competencies, strategic projects, and external means of action can be overcome if the actors are represented in the territory's prospective reflection committee, of which a work team is part. In a platform installed to continue advancing in this purpose, it will be easy to identify when an actor that has high control over a strategic challenge does not help in the

**Table 2.3.** Organization of the context analysis by phases and components (Valverde *et al* 2020).

| No. | Objectives | Components | Indicators |
|---|---|---|---|
| Axis 1. Green and eco-efficient environment | | | |
| 1 | Possess competencies regarding the availability, coverage, access, and use of services related to water in cities. | Water resources | 1. Water coverage<br>2. Efficiency in the use of water<br>3. Efficiency in water supply<br>4. Availability of water resources |
| 2 | Adequately monitor the emissions of suspended and dispersed particles in the air, as well as GHG. | Air | 1. Air-quality control<br>2. Concentration of pollutants in the air |
| 3 | Have eco-efficient and environmentally friendly energy alternatives available to meet the needs of society. | Energy | 1. Energy coverage<br>2. Energy efficiency<br>3. Alternative and renewable energy |
| 4 | Possess the skills required to study the availability of natural resources, including quality and quantity variables. | Ecosystems | 1. Availability<br>2. Access<br>3. Distribution<br>4. Use |
| Axis 2. Services, compact and integrated urban mobility | | | |
| 5 | Determine the ease and degree of safe access to established and emerging social and basic services in the urban and periurban sector. | Sanitation | 1. Urban sanitation coverage<br>2. Wastewater treatment<br>3. Drainage effectiveness<br>4. Collection coverage: solid waste<br>5. Final disposal of solid waste<br>6. Solid waste treatment |
| 6 | Determine inherent aspects of access, treatment, and control in health matters. | Health | 1. Health level<br>2. Provision of health services |
| 7 | Consider the achievement of medium and long-term objectives in education and influence the inclusion of minority sectors. | Education | 1. Educational quality<br>2. School attendance<br>3. Higher education |
| 8 | Establish competencies and attributes, guaranteeing the safety and wellbeing of society. | Security | 1. Degrees of citizen violence<br>2. Citizen confidence in security matters |
| 9 | Determine the transportation problem, agreeing on spaces for conciliation and mutual agreements for the benefit of society. | Mobility and transportation | 1. Balanced transport infrastructure<br>2. Road safety<br>3. Degrees of traffic congestion<br>4. Balanced demand and supply<br>5. Public transport planning |

| | | | |
|---|---|---|---|
| Axis 3. Solid, innovative, and competitive economy | | | |
| 10 | Contrast, tabulate, and study the existing biases in periurban sectors with greater emphasis. | Urban inequality | 1. Poverty, migration, and marginalization<br>2. Gender and generational age<br>3. Income inequality |
| 11 | Study the development of formal and informal trade and evaluate the scope of small and medium enterprises (SMEs). | Employment | 1. Formal and informal employment<br>2. Unemployment<br>3. Productive growth of SMEs<br>4. Competitive business development |
| Axis 4. A safe city and prepared for risk | | | |
| 12 | Plan land use and occupation of territory in the study cities. | Land use | 1. Density<br>2. Housing<br>3. Green and recreation areas<br>4. Food safety |
| 13 | Determine the efficiency and effectiveness of the use of resources for various productive economic purposes, as part of a comprehensive assessment of the environment—territory. | Territory occupation | 1. Productivity<br>2. Degradation<br>3. Ecosystem carrying capacity<br>4. Economically active population<br>5. Sustainable livelihoods<br>6. Productive entrepreneurship<br>7. Alternative production and value chains |
| 14 | Determine the degree of risk resulting from the physical and anthropic nature of the study cities. | Risk management | 1. Climate threats: extreme events<br>2. Socio-environmental vulnerability<br>3. Risks of natural disasters |
| Axis 5. Empowered, inclusive citizenship with social equity | | | |
| 15 | Promote management, participation, and decision-making in response to the needs of a biocultural, sociopolitical, and productive economic order, considering the appropriate competencies for management. | Governance | 1. Key public management functions<br>2. Service delivery<br>3. Citizen participation in public management<br>4. Access to public information<br>5. Sociopolitical representativeness |

(*Continued*)

**Table 2.3.** (*Continued*)

| No. | Objectives | Components | Indicators |
|---|---|---|---|
| 16 | Achieve cooperation in government interactions with civil society, in relation to the mutually agreed interaction of humanitarian, social, and environmental aid. | Governability | 1. Accountability to citizens<br>2. Social control of public management<br>3. Transparency and prevention of corruption |

achievement of an associated or collective objective that contributes to overcoming the collectively constructed challenge, which is why it follows that his positioning, in principle, is contrary to that objective, no matter how much he tries to tell us otherwise.

### *2.3.4.2 The participatory diagnosis*

To carry out the participatory diagnosis, the results of the context analysis are considered, which provide the necessary inputs to continue addressing the consistent elements of the study, but this time, for the planning and implementation of a series of prospective workshops that promote greater quantitative and qualitative precision regarding the state of affairs of the citiy under study regarding the issues raised. Consequently, table 2.4 shows how the participatory diagnosis considers the application of comprehensive workshops using the Prospective Workshop technological tool, the exercise of which reveals a quantitative and qualitative agreement of common and individualized perceptions of the best weighted attributes.

Based on the objectives of the Context Analysis, in this exercise 36 attributes are added, ordered according to the component and axis to which they correspond. It is necessary to clarify in this instance that the attributes are criteria exclusively associated with institutional management issues; that is to say, of organizational and financial administration, as well as of technical and operational management, and that the entities summoned for this exercise follow, for which they are 'dragged' from the previous stage. Therefore, the structure considers the same 5 axes, 16 components, and 36 attributes that synthesize and replace the indicators to be evaluated by specialists, which are obtained from the methodological documentation for cities in transition.

During the two sessions of each workshop unit—one for induction and one for voting—the attributes defined by the work team are exposed to the participants so that they can be adjusted, complemented, or eliminated. In conclusion, the participatory diagnosis must be two-dimensional, which implies that consistency is required between the starting information for each city (which we can call the baseline) and the information resulting from the application of the workshops, in order to obtain the perception of the participants, and facilitate the global vision of the problems that the city faces. This requires a broad exploration of the consequences of changes on activities, especially ecological, economic, and cultural, which must consider contrasted and different points of view.

**Table 2.4.** Elements for participatory analysis (Valverde *et al* 2020).

| Axis | Component | Attribute | Description |
|---|---|---|---|
| Green and eco-efficient environment | Water resources | Regulation | Projection and updating of the normative regulation regarding the protection and safeguarding of water resources, directing their availability, access, and use in the urban context. |
| | | Characterization and protection | Identification and updating of the inventory of water bodies according to their origin and their type of use, for the proposal of policies, strategies, and binding actions between the integrated water resourced management and urban integrated watershed management. |
| | | Environmental function | Potential consideration of the ecological flow to analyze the water balance in urban ecosystems, establishing the balance of its environmental function with respect to its protection, conservation, and use. |
| | Air | Regulation | Regulatory approach for the prevention, mitigation, and control of pollutant particle emissions and greenhouse gases, as well as the reduction of noise pollution in the city. |
| | | Characterization and control | Updating of information systems related to the inventory, monitoring, and control of polluting particles and greenhouse gases in the city. |
| | Energy | Regulation | Normative approach for the promotion of alternative energies, based on the creation of tax incentives that motivate the use of clean and eco-efficient technologies in the city. |
| | | Characterization | Determination and updating of the inventory of renewable and non-renewable energy sources; the availability of eco-efficient and environmentally friendly energies to meet the needs of the urban population. |
| | | Production | Development of projects and initiatives that promote the use of clean and renewable energies for eco-efficient production that meets the energy needs of the city. |
| | Ecosystems | Zoning and characterization | Registry and inventory of protection and conservation areas that fulfill the environmental function of recreation, transition, and integral use in the city. |

*(Continued)*

**Table 2.4.** (*Continued*)

| Axis | Component | Attribute | Description |
|---|---|---|---|
| | | Environmental function | Determination of the eco-geographic and jurisdictional connection of the urban space and territory, due to its climatic, biocultural, sociopolitical, and economic–productive implications. |
| | | Revaluation | Promotion and institutional organizational—financial and technical—operational strengthening, aimed at the management of protection and conservation areas, which promote effective actions in the urban environment. |
| Services, compact and integrated urban mobility | Basic sanitation | Infrastructure and equipment | Construction, improvement, and implementation of drinking water supply systems, as well as urban wastewater and solid waste treatment systems. |
| | | Services | Optimization of the provision of emerging basic services established in the urban and periurban sector, promoting management, management alternatives, and treatment alternatives with positive and viable effects in the city. |
| | Health | Infrastructure and equipment | Improvement of the hospital infrastructure at all levels, with the incorporation of medical specialties that meet the established demand. |
| | | Services | Implementation of a quality control system in health services, considering coverage and access to medical specialties, according to demand. |
| | Education | Infrastructure and equipment | Expansion and improvement of the educational infrastructure at all levels, in order to meet the established demand. |
| | | Services | Implementation of a quality control system in education services, considering coverage and access to regular, higher, alternative, and special education, according to demand. |
| | Security | Infrastructure and equipment | Installation, adaptation, and improvement of urban security surveillance systems, incorporating suitable technological equipment for the control of illegal activities, road accidents, and the risk of violence. |
| | | Services | Projection of the integrity, security, and wellbeing of society, coordinating citizen security action plans with the participation of the municipal guard and the inhabitants of each district in the city. |

| | | | |
|---|---|---|---|
| | Mobility and transportation | Infrastructure and equipment | Expansion and improvement of the road infrastructure, in order to meet the established demand, in accordance with the projections for the mass and integrated transportation systems. |
| | | Services | Improvement of road mobility of urban transport systems to generate mutual benefit between users and service providers in the city. |
| | | Integration | Promotion of an urban public transport system to improve links with expansion areas in the city. |
| Solid, innovative, and competitive economy | Urban inequality | Social inclusion | Reduction of existing biases of social inequality in the most vulnerable groups, optimizing the conditions of access to basic and social services in the urban area. |
| | | Life quality | Promotion of values that encourage effective coexistence, preserving the identity of the city's natural and cultural heritage. |
| | Urban inequality | Production planning | Determination of the efficiency and effectiveness of the use of resources for various economic purposes, according to established production planning, taking advantage of the potential of the city. |
| | | Employment supply and demand | Determination of job opportunities for new generations of professionals according to personal and collective competencies, capacities, values, and aptitudes. |
| A safe city, prepared for risk | Land use territory occupation | Regulation | Normative and technical regulation of land use in accordance with the geomorphological and environmental conditions of the city. |
| | | Zoning and characterization | Register and inventory regarding the aptitudes and land uses according to the ecosystem carrying capacity and the productive potentialities in the city. |
| | Risk management land use | Regulation | Normative and technical regulation of uncontrolled settlements according to urban use, distributed throughout the city. |
| | | Management competences | Promotion of a territorial planning system that integrates platforms with an urban municipal perspective and is of strict regulatory compliance with respect to urban settlements. |
| | Territory occupation | Regulation | Normative and technical regulation regarding the probabilities of prevention and control of risk events likely to originate in the city for natural, climatic, or anthropic reasons. |

(*Continued*)

**Table 2.4.** *(Continued)*

| Axis | Component | Attribute | Description |
|---|---|---|---|
| | | Management competences | Promotion of a risk management system that integrates platforms and observatories aimed at the prevention, mitigation, and control of climatic and natural threats and environmental vulnerabilities. |
| Empowered, inclusive citizenship with social equity | Governance | Administration | Strengthening of organizational—financial management, participation, and decision-making in response to the biocultural, sociopolitical, and economic–productive needs of the city. |
| | | Management | Strengthening of technical and operational management, and of strategic planning in response to the biocultural, sociopolitical, and economic–productive needs of the city. |
| | Governability | Accountability | Promotion of spaces for social interaction between the governmental and private levels, in a relationship of mutual reciprocity and democracy within the city. |
| | | Social control | Promotion of transparency and accountability mechanisms to improve the evaluation of public management. |

Finally, the application of this technique and its respective tool can be updated as more players join, the scales are expanded, some productive sectors that are decisive for the economy of a city are emphasized, and disciplines are added to the situational analysis of each territory study, which implies working on a network platform without restrictions of time or place, which facilitates group work that produces immediate results, where preconceived ideas are eliminated and a dialogue is built that allows an understanding of the definitions established in order to carry out a vote that supports the stated objective.

*2.3.4.3 The purpose of priorities in the selection of actions*

In this stage, discussion groups are used, in which structural criteria are no longer evaluated, as in the case of MACTOR or participatory diagnosis (with the application of the Prospective Workshop), but, on the contrary, it is about the collective evaluation of a series of functional measures, in order to project a series of emergent actions for the city studied. In this way, its consideration allows us to understand whether the directionality of the cities in their transition process prefers short-term or long-term scenarios.

The prioritization of strategic measures is perhaps the most important task, since it conditions the work and action of all the results that have been found up to this stage, which correspond to the planning and projection of the proposed prospective study. The process of prioritization requires that the central objective of the project, the underlying goals, and the strongest interventions be specified, based on policies and multicriteria actions, which allow the behavior of the object of study to be changed by all the actors involved, in two ways:

- Reactive: making decisions as problems or needs appear without differentiating the characteristics of the locations where they originate. This is a consideration of the short term and a large space.
- Proactive: creating alternatives and acting to the extent that there are things to solve, but in a programmatic way and distinguishing the locations with the greatest vulnerability, the presence of socio-environmental problems, and conflicts. This is long-term thinking for a specific space.

For this, table 2.5 shows the selection as the beginning of the process of control of the (36) variables and strategic measures that replace the (36) management attributes and that are fundamental to understand the direction that a city adopts in its transition process, maintaining the axes and components of the previous stages obtained from the methodological document of cities in transition:

The exercise starts from the configuration of the results obtained with the participation of specialists consulted and determined by the Color Insight tool, which shows us a projection that evolves towards precise considerations in each of the five determined axes. We introduce both the results of MACTOR and the Prospective Workshop to this context, consequently assessing the perception of the probability and desirability of activating certain strategic measures that are ultimately conceived as purely operational decisions.

**Table 2.5.** Elements for the initial determination of strategic measures (Valverde *et al* 2020).

| Axis | Components | Variables | Strategic measures |
|---|---|---|---|
| Green and eco-efficient environment | Water resources | Quantity | System of availability, coverage, access, and use of services related to water in the city. |
| | | Quality | Monitoring of water for its optimal use and treatment, according to the type of use. |
| | | Usage practices | Responsible environmental practices for the proper use of water. |
| | Air | Quality | Air-quality monitoring, accounting for emissions of particles and greenhouse gases for their reduction or attenuation. |
| | | Acoustic impact | Monitoring of noise pollution to reduce or mitigate its impact on human health. |
| | Energy | Access and coverage | Assembly of renewable energy systems using eco-efficient technologies for the promotion and distribution of clean energy. |
| | | Cost and efficiency | Incentives for the assembly of low-cost renewable energy systems, integrated into the city's architectural infrastructure. |
| | Ecosystems | Degradation | Comprehensive evaluation strategy for impacts, damage, and effects in urban socio-natural systems. |
| | | Restoration | Multicriteria strategy for urban recovery and the revitalization of degraded areas; implementation of an early warning system for the care of areas of high protection and conservation value. |
| Services, compact and integrated urban mobility | Basic sanitation | Availability, access, coverage and distribution | Attention and response system regarding the provision of services related to basic sanitation: drinking water, sewerage, and solid waste management. |
| | | Treatment | Effective treatment systems for the mitigation and control of pollutant loads that can affect human health. |
| | | Cost and efficiency | Incentive program and regulation of fees for the responsible use of basic services and the promotion of good practices to minimize the environmental impact of waste and residues. |
| | Health | Access, coverage, and attention | Policies and mechanisms for access to a more efficient and effective service with the incorporation of different medical specialties and care for the most vulnerable sectors of the city. |

| | | | |
|---|---|---|---|
| | | Communication | Awareness and education campaigns for inclusive and high-quality health, aimed at preventing disease and generating healthy hygiene and eating habits in the urban context. |
| | Education | Training | Cross-training programs incorporating environmental issues aimed at promoting responsible values and attitudes towards the natural and social environment of the city. |
| | | Scientific research | Comprehensive research programs aimed at practical scientific and technological answers and solutions to environmental issues, encouraging more conducive alternatives for urban development. |
| | | Social extension | Scientific and technological innovation programs that encourage the biocultural, sociopolitical, and economic–productive transformation of the urban environment based on mobilization, exchange, and financing strategies for new researchers. |
| | Security | Attention to violence | Comprehensive plans to promote effective citizen security actions, identifying risk factors for violence in the city. |
| | | Social welfare | Participatory strategies for the promotion of actions aimed at peaceful coexistence in the community in all districts and neighborhoods of the city. |
| | Mobility and transportation | Road planning | Mass and integrated transport systems to reach coverage and connection of central and peripheral areas in the city. |
| | | Vital education | Orientation and road education campaigns for users and transport service providers to provide a better service in compliance with traffic regulations. |
| | | Road safety | Policies and regulatory provisions to guarantee the road safety of pedestrians and users of alternative means of transport (motorcycles, bicycles, others). |
| Solid, innovative, and competitive economy | Urban inequality | Poverty and marginality | Studies to enhance the dynamics of influence; urban–rural dependence that allows encouragement of the egalitarian provision of services, especially in vulnerable areas of the city. |
| | | Gender and generationality | Comprehensive strategies to promote respect for human rights in the consolidation of new urban centers with the participation of women and youth from the city. |

(*Continued*)

**Table 2.5.** (*Continued*)

| Axis | Components | Variables | Strategic measures |
|---|---|---|---|
| | Urban inequality | Productive enterprises | Promotion of seed funds for the development of formal productive and commercial enterprises in the city. |
| | | Incentives to production | Financing programs for SMEs that promote environmentally responsible productive activities. |
| A safe city, prepared for risk | Land use/territory occupation | Green and recreational areas | Municipal tree planting and maintenance program for areas conducive to enjoyment of the urban landscape. |
| | | Urban areas | Urban planning system consistent with defined land uses and aimed at restructuring and expanding the requirements for new developments. |
| | Risk management/ land use | Migratory settlements | Municipal platform for territorial regulation for the availability, sanitation, and urbanization of new properly planned areas. |
| | | Gentrification | Strategies to reduce the impact of displacement which minimize invasive settlements in traditional sectors and neighborhoods of the city. |
| | Territory occupation | Threats | Protocols for the identification, prevention, and control of threats from probabilistic occurrences of extreme weather events. |
| | | Vulnerability | Protocols for determining the degrees of exposure, sensitivity, and adaptive capacity of urban socio-natural systems. |
| Empowered, inclusive citizenship with social equity | Governance | Representativeness | Mechanisms of citizen participation for the deliberation of policies, strategies, and actions related to the urban context. |
| | | Receptivity | Mechanisms for the promotion of strategies and actions of arbitration, mediation, and consensus among all the actors involved who have competencies in urban issues. |
| | Governability | Stability and transparency | Permanent attention system of timely and sufficient information as a mechanism for accountability and social control. |
| | | Efficiency and legitimacy | The institution of scientific and technological citizen consultation councils for co-managed and effective decision-making for the city. |

By replacing the attributes with variables of a more technical and operational nature, it is possible that some components not prioritized by the actors in the previous exercise may, at this stage, become important and come to the fore, since they are strategic measures accompanied by a choice in matters of time and space; consequently, the results obtained will emerge from the positions reflected (as detailed below) by the vote of the consultation group.

The group of specialists determines strategic measures that consider measures for the case of proactive responses aimed at sustainability, development, and resilience issues, as well as for the case of reactive and emerging care, linked to issues of economic sustainability, growth, and adaptation/mitigation.

#### *2.3.4.4 Multicriteria interventions*

The last exercise of a prospective study is the projection of scenarios from the set of multicriteria interventions (based on policies, attributes, and strategic measures) that are carried over from the previous exercises (MACTOR, the Prospective Workshop, and Color Insight), which is carried out to create a comparative judgment between heterogeneous measures and evaluate the capacity of various actions in order to consolidate a broader program to be executed by those responsible and the beneficiaries.

This will allow a discussion of the content of the measures that urban programs (or plans) incorporate, if they are conceived from a scientific point of view and if these measures were statistically determined; the common legal and political framework that supports and supports them is also analyzed, and finally, the allocations of material, human, and economic resources (not contemplated in this study) necessary to implement them are discussed for the two aforementioned scenarios: that of the short term, which grants reactive attention in large spaces; and the long term, which provides proactive responses in localized spaces.

Based on these considerations, the criteria resulting from the previous exercises are established, which will be evaluated by the work team, grouped into:

- Objectives, which were taken into consideration in the Context Analysis.
- Attributes, which were assessed in the Participatory Assessment.
- Measures, which were prioritized in the previous step of Strategic Measures.

Consequently, it is at this stage that the concepts acquire another description and meaning for each study city, since, as we discussed previously, each of the study cities has totally different urban, environmental, and institutional management characteristics, not to mention the geographical and even climatic considerations that differentiate their spatial and territorial correspondences.

Multicriteria interventions (which are the product of the sum of policies + strategic measures), are determined by applying software called MULTIPOL, which constitutes a prospective tool that allows the evaluation of multicriteria interventions, in order to compare them and establish a weighting that facilitates technical and political decision-making by the related actors of each city. Therefore, multicriteria elements are integrated and policy approaches are postulated, supported by the objectives defined at the beginning, and corroborated by the constituent items of each attribute contributed by the components of the project, as a way to evaluate the fulfillment of the emerging strategic measures and the priorities for each city, so the comparisons are determined by following the best valued weightings.

## 2.4 Conclusions

A prospective study to try to drive the 'urban transition', should generate a series of reflections on the decision-making capacity of urban institutional actors to guide the direction of their own trends, modulate the consequences of their actions, and graduate the predictable impacts of disasters which could predictably affect their city in the projected scenarios. It is a question of inviting responses as multicriteria interventions, in the face of all the biocultural, sociopolitical, and economic–productive problems and needs that constitute the environment in which the city unfolds.

The process begins with a heartfelt concern, that of feeling the need to respond to a transitional phase. While it is true that development is change, and this is a constant of territorial systems, determining one phase of it as a transition requires the objectification of a starting point and a destination, which acts, at least, as a guiding hypothesis for action. Once the city has an internal objective and a trajectory in its external environment whose perceptions are known, two non-limiting and almost complementary approaches are considered: the approach to economic sustainability, growth, and adaptation/mitigation and the approach to sustainability, development, and resilience. Each city manages its capabilities according to a temporary game oriented by institutional (actors), structural (political), and functional (actions) criteria.

The prospective study objectively approaches the definition of behavior in terms of the state of a city, but the greater usefulness of this type of study is obtained from self-reflection and the understanding by the actors of their abilities to execute the policies and actions they want. It is important that there is a cognitive and temporal lag in the perceptions of those involved, which causes their decisions to vary during the process of making them, introducing distortions and a possible zigzag configuration in the composition of the two transitions as defined in their purest states, which tend to 'denature' into original solutions which define their own models, which is a constant in geographical relationships and ensures the diversity of places.

On a small comparison scale with a high degree of abstraction, we could well say that in the three Bolivian cities where the proposed urban territorial project logic was applied, there was a strong willingness to start the transition to sustainability, development, and resilience. Mieres showed a desire to adapt to a new scenario, one that was unknown and subject to high uncertainty due to the dwindling internal demographic potential to facilitate urban regeneration, by setting up a city model that is not as intensely productive as it was in the immediate past, one that takes advantage of the opportunity to re-qualify, orienting itself to aspects previously neglected by systemic requirements, such as mobility, safety, energy efficiency, landscape recomposition, and better management of public space.

Tacking the ordinary urban conflicts of territorial management while addressing new environmental challenges is a whole; these are not just isolated things. Fixing the city is not a matter of re-engineering. The answer on both planes is joint and involves tackling the capabilities of cities to redesign and manage their urban policies using a holistic approach, in order to address systemic challenges in advance and prepare to react promptly to the urban surprise that, by definition, occurs in any interface or transition period. The current transition seems to lead into the metropolitan century, which requires constant proactive adjustments, intelligent perspective, and a comparative approach.

As is known, there are many versions and nuances to defining the city model that we are looking for, since each city responds in a different way to its own geographical, ecosystem, cultural, and other characteristics; however, it is possible to define general and specific characteristics or criteria that are the object of this proposal, which, based on the perceptions of different stakeholders (mostly specialists), focuses on: (a) mapping actors, roles, and interests, (b) systematizing the starting information obtained from the prospective workshops, and (c) ranking, based on the prioritization results, the actions that lead the study cities to be conceived as model spaces—territories of operational and orderly urban planning.

Having carried out a first trial of the project exclusively in Bolivian cities, with interesting results, it has been verified that the prospective workshop has been an essential tool that allows us to determine the future aspects that contribute to the scope of a city in transition, but, above all, infer which actions would make it operational. However, for a complementary study to be more effective and channel the preliminary hypotheses requires the articulation of greater actors that are part of the urban, environmental, and institutional development of a city, in different scenarios, and considering broad temporal and spatial possibilities.

This articulation must be seen as a process by which permanent relationships are established between the agents of a territory, based on the search for common objectives that transcend particular or sectoral interests, without negating them, although placed in a cooperative situation for socio-environmental common well-being. Therefore, this implies the identification of the interest or the common good, which can be achieved based on negotiation instances which establish the rules of the game, within a framework of power relations that allow a certain flexibility and can change over time. In this sense, it is intended to improve the conditions to create spaces that allow the actors involved to be agents of change for their own collective well-being in the community.

There is no doubt that the prospective workshops carried out in the initial study cities in Bolivia deepened the understanding of the urban, environmental, and institutional requirements of macro-districts, districts, areas, or more specific neighborhoods, since addressing these issues has always been a demanding issue. For this reason, it is important to know the voices of the main stakeholders, the priority aspects, and the consensus on certain responses, with the intention of achieving more specific actions that directly impact the problem and that take into account the perception of the inhabitants of the area, as well as the technical and strategic views, and those of the decision-makers of urban spaces and territories.

A possible proposal that would complement the preliminary study in Spanish cities could establish, in a common way, the possible actions and policies that need to be addressed and projected, and that remain to be incorporated into territorial planning, not from an institutionalized perspective, but rather from an operational perspective. The project, in its geographical environment, will train, guide, and agree collaborative relationships with the true stakeholders and beneficiaries, in time to evaluate the contribution of each actor in the achievement of objectives, so that they know how to differentiate and react to the conceptual categories that have been assessed in the first study: sustainability vs. economic sustainability, development vs. growth, resilience vs. adaptation/mitigation. This situation must be accompanied by a space in which to share the scope of the training and discuss experiences, as well as consideration of a collaboration between those responsible for different cities.

## Additional reading

Astigarraga (2016); Bas (2002); Godet (2000); Haase *et al* (2016); Martin and Valverde (2020); Rodríguez (2016); Rodríguez and Menéndez (2019); Valverde (2019); Valverde (2020)

## Bibliography


Astigarraga E 2016 Prospectiva Estratégica: orígenes, conceptos clave e introducción a su práctica *Revista Centroamericana de Administración Pública* **71** 13–29

Bas E 2002 *Prospectiva: Cómo usar el pensamiento sobre el futuro* 2nd edn (Barcelona: Ariel)

Godet M *et al* 2000 *Caja de herramientas de la prospectiva estratégica* Notebook published by Gerpa in collaboration with Electricité de France, Mission Prospective Cuaderno 5 (4th edn actualizada)

Haase A, Bernt M, Großmann K, Mykhnenko V and Rink D 2016 Varieties of shrinkage in European cities *European Urban and Regional Studies* **23** 86–102

Martin D and Valverde A 2020 *Ciudades en transición: Estudio prospectivo de análisis multicriterial en Bolivia* (Integra Educativa. Ciudades en Transición) vol XII p 36

Rodríguez F 2016 (dir.). *Sustainable and Integrated Urban Development Strategy*. Mieres Project. Selected in the 2016 call. EDUSI-Mieres. Resolution 09/29/2016 of the Secretary of State for Budgets and Expenditures, 1st Call for the selection of strategies for sustainable and integrated urban development, co-financed by FEDER 2014-10, Order HAP/2427/2015, of 11/13

Rodríguez Gutiérrez F and Menéndez Fernández R 2019 Contracción territorial y ordenación urbana para encarar el ciclo post-minero en Mieres (Asturias-España) *Ciudades* **22** 99–122 *Contracción territorial y ordenación urbana para encarar el ciclo post-minero en Mieres* (Ciudades: Asturias-España)

Valverde A 2019 *Técnicas de prospectiva aplicadas a la investigación científica* (Integra Educativa) p 35 Transgresiones científicas en investigación posgradual vol XII no 2 pp 49–66

Valverde A 2020 *Prospectiva Territorial Urbana. Una estrategia de indagación y priorización para accionar intervenciones multicriterio en ciudades en transición* (Integra Educativa.Ciudades en Transición) vol XII no 3 p 36

Valverde A, Rodríguez F, Wrenke A, Curras T, Ramos P, Johnson B and Martin D 2020 *Strengthening Citizen Platforms: Sustainable and Resilient Cities* (UCB, WWF, Swedish Cooperation, 2019–2020) (La Paz: Sverige Embassy and World Wildlife Fund)

# Chapter 3

# An urgent urban design challenge: time, sustainable mobility, and the humanization of collective space

**Miguel Bandeira, Filipa Corais and Fátima Pereira**

This chapter is intended as a reflection on three concepts that are considered interdependent: time, sustainable mobility, and the humanization of the collective space. The identity and position of each city in relation to the level of sustainable urban mobility achieved and the appetite for the humanization of its collective spaces are determined by its location in time, with regard to its degree of maturity in the implementation of these concepts. It is therefore crucial to define them and to determine mechanisms for critical evaluation and monitoring in time that contribute to the definition of standards and a theory that helps to accelerate the processes of changing behaviour and paradigms in terms of mobility and urban design.

## 3.1 Time and sustainable urban mobility

The concept of time is inherent in the forms the city contains; each form corresponds to a time. Time is also an intrinsic component of mobility and its project design. The dynamics prevailing in the temporal contexts of the history of architecture and urbanism, which have conditioned the present situation, as well as those that are projected towards a future paradigmatic destination (which will not be as far away as some would like to believe) determine these temporal idiosyncrasies that shape urban mobility. This 'crack in time' (Silva 2014, p 247) cannot always be perceived accurately, but the (un)certainty of the future positions it in the present, not neglecting the past. Awareness of this 'folded time' (Silva 2014) is fundamental to the conjugation of theory and practice through the incorporation of the dynamics of spatial transformation with the desired changes in social behaviour. In this respect, the project, although aware of the goal it intends to achieve, may incorporate some purposeful and camouflaged programmatic vagueness that imprints the flexibility required to accompany and adapt to the temporal evolution of these concepts.

doi:10.1088/978-0-7503-3971-1ch3

This (patient) ability to incorporate the phasing of objectives into urban design, combined with the innovation component, determine the (degree of maturity of the) role of the architect and the planner in their act of conducting the mobility planning process (Corais 2021, p 27).

Time is the determining factor in the completion of sustainable urban mobility in a given territory or community that is gauging its resilience in the face of imponderables, such as the current pandemic crisis. Given the positive effect of urban acupuncture measures in this area, their effectiveness depends not only on their strategic location, the scale of intervention (small but energetic), or the degree of collective involvement, but also on the speed of action. Time is therefore a decisive variable in achieving the desired catalytic effect.

The uneven position of cities in Northern Europe, in terms of their maturity in relation to sustainable urban mobility invokes a time not synchronized with Western European cities and awakens knowledge of the solutions found in these cities, by analogy with the problems being experienced in the latter. This knowledge of an admissible solution, according to Corboz, even if not applicable to the model in a direct way, due to its inherent specificities, offers utility in the conception of the process of finding an alternative solution (Corais 2021, p 91).

Mobility is an area that depends on human action; following the example of the progress of other countries in this field, it is possible today to glimpse a vision of the future, without compromising the transition from the past (very deep-rooted) to the present.

## 3.2 Sustainable urban mobility and the humanization of collective space

Sustainable urban mobility, in addition to its inherent economic and environmental benefits (according to the demanding criteria of urban design), contributes to boosting the humanization of collective space. On the other hand, urban design is a determining tool for the promotion of sustainable urban mobility, without which, the desired results will not be achieved. In this sense, it is considered important to value the reciprocal relationship between urban design and sustainable mobility, which has been so neglected in some cases (Corais 2021, p 21). As a result of this interaction between urban design and sustainable mobility, the collective space is given the value of sustainability that integrates economic, social, and environmental benefits.

While the concept of urban design has long been present in the consciousness of architects and urban planners, its inclusion of methodologies that promote sustainable urban mobility has not always been and is not always considered, even today. This discrepancy and alienation between urban design and sustainable mobility contradicts the programmed and desired cycle of the 'path' to be followed until the goals set for 2050 in the Paris Agreement are achieved. In addition to these temporal consequences in relation to the achievement of sustainability targets, there are also effects on the humanization deficit of collective spaces and on people's quality of life.

This topic has become more important in the context of the current COVID-19 pandemic, although this is not the first crisis that has made us think about the

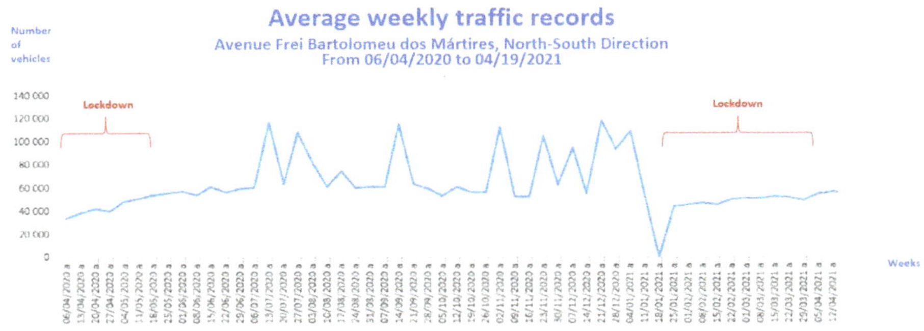


**Figure 3.1.** Graph of the average number of vehicles per week circulating in Avenue Frei Bartolomeu dos Mártires in Braga, since the beginning of the COVID-19 pandemic. Source: CMB 2021 Municipal Traffic counts, Braga.

impacts of urban planning and design on public health (Honey-Rosés *et al* 2020). Actually, in the times of lockdown, there was a reduction in car traffic (of about 50%) that resulted in a significant reduction in air and noise pollution (figure 3.1). This situation may lead to greater awareness among the population and the administration regarding these matters and may contribute to the consideration of mobility models and standards for the future.

The main prevention policies related to COVID-19 in public spaces focus on use restrictions (which are evident in the images of empty streets and parks publicised in many cities around the world—which have turned into ghost towns) and the establishment of social distancing. On the other hand, as opposed to the interdiction of the usufruct of public spaces (e.g. fenced playgrounds and park benches) there is a growing propensity to use them when allowed, defining new opportunities for micro mobility, which may imply the need for the redistribution of public space in favour of more sustainable modes of transport (Honey-Rosés *et al* 2020).

Again, in this pandemic context, the balance and fairness of the use of public space capacity were obvious as, according to Valentino-Devries *et al* (2020), it was the lower-income workers who were compelled to use more public space in their commuting in this period, as the higher-income jobs made more use of teleworking.

In Braga, as in the rest of the country, the general context of mobilisation in the fight against the pandemic mobilised the limited resources available, mainly in the enactment of emergency measures linked directly to public health.

In the field of mobility, initiatives were more oriented towards supporting urban transport than towards the possibility of starting building works in public spaces, which in Portugal require tight and time-consuming public procurement rules (including pop-up measures). As a proposal, in the future, Portugal will have to consider agile intervention measures that are adequate to the contingency situation we are living in. However, this reality will have to be reflected upon and approved at the national level.

Nevertheless, right after the period of lockdown, Braga closed an important street to traffic in its historical centre (Rua Gonçalo Pereira), near the pedestrian area, for the installation of café and restaurant terraces, guaranteeing social distancing, urban living, and the economic dynamism of those businesses.

In May 2020 (at the end of the first lockdown period in Portugal) we conducted a survey of the population regarding the 'role of public space for physical and psychological wellbeing during a pandemic such as Covid-19', which attracted around 100 participants and was distributed online and carried out by Google Forms. Of the people surveyed, around 48% were residents in Braga and around 50% lived in apartments in the city centre; most were aged between 30 and 50 years old and belonged to a household of three people (32%) or four people (35%). In terms of the types of work carried out during the first period of lockdown, it was found that around 63% were teleworkers, around 36% had face-to-face work, and around 26% were fired or had their work suspended. In this survey, it was possible to see that the predominant mode of transport during the first period of lockdown was the car (in 58% of cases, always by car) and that the majority avoided public transport (96.5%). The bicycle was not an option either; 74.2% of the surveyed people never used this mode of transport and only 13.5% used the pedestrian mode of transport every time they went out. When questioned about the need for the implementation of intervention measures in the public space and the promotion of sustainable mobility that would contribute to the physical and psychological wellbeing of the population, approximately 75% answered affirmatively, and the same percentage considered that the main change should consist in the creation of wider pavements, 54% believed that more bicycle paths should be created, and 38% thought that more plazas should be created. Approximately half of the surveyed persons confirmed that they would be afraid of using public transport after the period of lockdown. Finally, when asked about their willingness to use more sustainable modes of transport when society returns to normality, 36.7% replied that they would be willing to change to the pedestrian mode, 18.4% to cycling, and 12.2% to public transport.

The relevance of sustainable urban mobility and decarbonisation is therefore essential, regardless of the commitments made for 2050. Indeed, cities are in a process of population growth, with consequences for accessibility, mobility, and quality of life, which require the adoption of sustainable non-'canonical' solutions. In this sense, in Portugal, the role of architecture and urbanism in the interdisciplinary component of mobility policies has been neglected, in both the theoretical and the practical domains, and it is therefore urgent to assume it as structural. This link between architecture, urbanism, and mobility, complemented by other inseparable disciplines, is based on principles of spatial organisation at the level of the *system of collective spaces, design of the urban form, and shaping the urban morphology* of cities (Corais 2021, p 16).

Looking back, it can be seen that the growth of cities in the mid-19th century, driven by the Industrial Revolution, underlined the city–country dichotomy and put pressure on the population densities and construction rates of traditional cities by introducing new construction methods. This changed and conditioned the mobility and accessibility of populations.

The study of, and research into the implications of urbanism and the design of public space on human social life constituted, from the end of the 19th century (with Camillo Sitte (in 1889), and afterwards with Ebenezer Howard (in 1902), in 'Garden Cities of Tomorrow'), a concern that would later be addressed in a quite distinct way by Le Corbusier (in 1923, in 'Vers une architecture' and in 1933, in 'La Charte d'Athènes') and Rem Koolhaas (in 1995, S,M,L,XL).

At the beginning of the 20th century, there were two distinct types of response to population pressure in cities: on the one hand, the Dutch model, developed by the architect Hendrik Berlage, based on classical urban forms, and on the other, the new principles developed by modernism (Gehl and Svarre 2013, p 39).

The urban conformations that have their origin in the logic of functional zoning, defended in the modern urbanism promoted by Le Corbusier in 1928 in 'Urbanisme' and the 'Cité Radieuse', defended the division of the city into four basic functions (living, working, travel, and leisure) and persisted until the end of the 20th century.

Thus, the characteristic sinuosity of the medieval city, composed of streets and squares as spaces of social congregation, was succeeded, in modernism, by the cities of rectilinear mesh, where the automobile assumed a preponderant role in relation to the pedestrian, and where the human scale lost expression. These cities grew in height and spread their perimeters, increasing the distances to be travelled on a daily basis and causing a worsening of the need to make more journeys by individual transport. As a result, families owned more and more cars (in 2017, in Portugal, there were about 490 passenger vehicles per thousand inhabitants, which represents a substantial worsening compared to the situation seen in 1990, when there were only 185.2 vehicles per thousand inhabitants (Pordata 2018)).

However, according to Ascher (1998), this urban practice, which he classified as a 'Fordist' city, did not prioritize urban design and sacrificed urban morphology to the functioning of the car-dependent city. Jane Jacobs, in the 1960s, also opposed this vision of the city, defending the diversity of uses and urban living in public space.

The discussion of the implications of the form of the city in terms of the human experience has acquired different outlines over the following decades; the prevalent view today defends an inclusive, readable, continuous, and accessible urban design for all pedestrians, following the logic of the supremacy of the humanization of collective space. Jane Jacobs was a mentor, in the 1960s, of the awareness of these matters and, more recently, Jan Ghel has been a faithful successor and diffuser in the world, through the application of these theories in various cities.

At the same time, de Solá-Morales (2008) addressed the assumption of the dichotomy between public and private spaces in an essay on urbanism theory, and founded the epistemological prevalence of the concept of collective space in place of the former. According to this author, the supremacy of the public versus the private space, which emerged at the end of the 19th century and which has perpetuated until the present, is related to dated political–urban ideologies which embodied various typologies of forms and which have integrated various theoretical contents over time. Since the strategic opening of strong urban axes and squares in emblematic cities, in which the public space has sometimes functioned as a support for the architecture of the city, in the perspective defended by Aldo Rossi in the 1960s, it has sometimes represented the real protagonist in the intense urban competitiveness, passing through the consolidation of urban parks on a supramunicipal scale, up to the present claim of an equitable redistribution of space by the various modes of transport; there is a false spotlight on this consolidation of the city's form on dominance, when the core of the problem is functionality, i.e. the connection of the spatial logic of the urban system. In this regard, de Solà-Morales (2008)) attributes the importance of city spaces to their urban experiences and to the social interaction that triggers and enhances the intrinsic appetite for private spaces (such as public spaces) as a collective heritage. Paradoxically, it also refers to the individual or collective appropriation of certain public spaces of the city. Gehl (2011, p 125), on the other hand, although he generally assumes an emphasis on public space throughout his discourse, acknowledges the tendency for private spaces to move into collective use (e.g. commercial areas, shop arches, and others) (Corais 2021, p 21).

The emphasis is therefore given to the system of collective spaces, which encompasses and overcomes the system of public spaces as far as the humanization of spaces is concerned. The system of collective spaces plays an important role in the structuring and characterisation of urban identity (Corais 2016). In fact, collective spaces, as an integrated system, assume an important function in the characterisation of urban identity, insofar as they structure the city formally and functionally (Corais 2016). Since the system of collective spaces is the primary element of the urban form, it sustains the city model, enhances its spatial integration, and manipulates mobility.

On the other hand, the systematization in the theory of '*3D*' *(Diversity, Density and Design)*, presented by Cervero and Kockelman (1997) corroborates the thought that the built environment influences commuting. Urbanism thus begins to integrate mobility as a proper substance of the architectural project, disregarding the concept of the segregated road network of the zoning approach (de Solà-Morales 2008). This theory, of the *symbiosis between architecture, urbanism, and mobility*, is later tested and synthesised by de Solà-Morales (2008), who assigns to architecture the role of controlling the impact of flows in public spaces that stimulate social practices (Corais 2021, p 18).

In this respect, the territorial approaches based on urban acupuncture developed by Manuel de Solà-Morales (2008) and Jaime Lerner (2003), have a very provocative application in the field of mobility that deserves to be explored.

de Solá-Morales (2008, p 23) makes an analogy between the epidermis and the city, and states that the skin of the city is made up of the physical matter of buildings, streets, public spaces, and the movements that embody human physical contact with this matter, ensuring the consummation of urban experience. According to de Sóla-Morales (2008), urban acupuncture, like medical acupuncture, acts strategically at neuralgic points and determines a much more comprehensive and widespread capacity for action throughout the city (Corais 2021, p 29).

On the other hand, it is proposed that the organisation of these one-off, urban acupuncture interventions in the global system of the city as a whole may find refuge in the adoption of the principles of urban morphology and respect for the typologies of urban fabric that pre-exist in cities. In fact, it is considered fundamental to associate urban morphology with mobility networks, and what is proposed is that this articulation be made through the implementation of urban acupuncture projects (Lerner 2003; de Solà-Morales 2008) integrated into a system of collective spaces within the city.

Urban acupuncture projects thus oppose the megalomaniac interventions of the past, promoted by the State using a *top-down* strategy, and implemented without the involvement of the population, and which had less immediate effects than those recommended in this type of more specific and incisive project. Four key features of this new way of intervening in cities stand out: the micro scale (but integrated into a global perspective); results in the shortest possible time; the contagious and catalytic effect on other areas of the city; and the involvement of communities. They therefore constitute topical interventions, but ones whose impact is very broad, both as a result of the inclusion of the community in the decision-making process for innovative projects, and of the effect of emanating change in social behaviour and its subsequent replication in the territory (Corais 2021, p 30).

The impact of urban acupuncture projects can be assessed through the evaluation of individual projects, or by the interaction of an articulated set of tactical projects implemented in strategic areas of the city. In any case, in one situation or another, if the impetus to speed up transformation processes in the city is to be realised, it must be part of an integrated vision of the interventions as a group and their interaction with the city (Corais 2021, p 30). On the other hand, the components of population participation, community involvement, and the co-responsibility of all stakeholders are fundamental to these procedures, ensuring that they do not become *top-down* (or *bottom-up*, as is the case with tactical urbanism), but operate through new social cohesion practices that place all stakeholders on the same level and focus on the same purpose. At this level, it is worth exploring the transition experiences (Roorda *et al* 2014; Sengers *et al* 2019) which are derived from the methodologies of the transition economy and which envisage the application of participatory approach tools in order to accelerate transitions in favour of sustainable development (covering its economic, social, and environmental components). Among the main tools for putting these experiences into practice are the transition arenas applied in a nonlinear process, which includes procedures such as setting up a diversified transition team, empowering agents, strengthening the collective spirit, reflecting on analysis, setting up change scenarios, involvement, and continuous evaluation (Roorda *et al* 2014).

The fusion between sociology, architecture, and urbanism, as related to the concepts of sustainable urban mobility, results in an approach of human interaction with the physical structure conceived by urban design and an analysis of the resulting impacts in favour of a paradigm shift towards the harmonization of social relations in public spaces; in particular, today, the plan to promote the decarbonisation of cities.

In the evaluation of the impact of projects and actions on changing the behaviour of the population, the theories developed by Hillier and Hanson (2003) or Gehl (2011) stand out, because they apply urban morphology to urban planning and design, valuing the humanisation of spaces and the spatial complementarity of the various urban components (buildings, public space, and people) that structure the territory in an interactive complex system.

Gehl and Svarre (2013, p 3), in line with the criticism of modern urbanism undertaken by Jane Jacobs in the 1960s, William H. White in the 1980s, and

Christopher Alexander, advocates the humanisation of the city by reducing the scale of public spaces and reducing the presence of cars, favouring pedestrians and urban living. In the 1960s, these authors developed academic methodologies with which to increase knowledge about people's behaviour on the move and their use of public space, generating approaches that, at the end of the decade, began to have practical application in urbanism.

According to Gehl and Svarre (2013) the key to developing appealing spaces for collective enjoyment lies in knowledge about the behavior in the built space, that is, in the dedication to the study of life in the collective spaces of cities and in the respective perception of the factors that encourage or discourage this use with a focus on people. The interaction between human life and the urban form is imbued with a complexity that must be decoded. According to this author, the collective valorization of a space lies not only in the daily activities necessary for the population (e.g., commuting, commercial needs) that it integrates, but above all, in the alternative and relaxing activities that it generates and that reinforce its social and human dimension.

Bornat and Shaw (2019), using children as key figures in the urban design of neighbourhood units, determined a methodology for involving the population that can be passed on to other groups in society. Although children are key players in mobility policies and the main agents of behavioural change in society, other vulnerable groups (e.g. people with reduced mobility and the elderly, among others) should not, however, be neglected.

A tool advocated by several authors (e.g. Ghel; Bornat and Shaw) is to learn through observation how the population perceives and enjoys collective spaces in their daily lives, and what factors influence their behaviour in those spaces. These methodologies of incorporation of the population's involvement in the analysis and enquiry about the population's role in urban design and in the use of collective space take different forms, such as the use of photographic recordings, group walking tours, the mapping of routes and spaces or the relationships between spaces, time, and activity records, the application of interviews and surveys, the collective classification of spaces using photos or maps, and other procedures. These instruments assume the dual roles of, on the one hand, offering their users the opportunity to discover a new perspective on spaces that questions current uses and reinforces the sense of belonging and involvement and, on the other hand, demonstrating to the decision-maker the real collective needs that should be contemplated in urban design.

Spatial analysis combined with the analysis of the human component allows us to grasp collective perceptions of space experimentation and test the appetite of the citizens for the promotion of equitable and inclusive sustainable urban mobility; it is also a factor in promoting the quality of urban life (Corais 2021, p 32).

Awareness and the incorporation of social science tools in the fields of architecture and sustainable urban mobility planning, due to the interconnections generated, translate into an increased potential for the consolidation of change in the paradigm process (Corais 2021, p 31). Thus, the interdisciplinary articulation of these theories using the architecture of mobility, promoting spatial humanization, and stimulating the adoption of good practices in human behaviour, advances the achievement of the goals set for 2050 and promotes the implementation of preventive measures to inhibit the spread of disease, such as COVID-19, ensuring social distancing when travelling.

Hillier and Hanson (2003, pp 24–6) highlight the importance of social interaction that proves that the human presence in spaces catalyzes other social experiences in a cyclical and reciprocal way and that this awareness of human presence is determined by architecture. The authors thus defend the interaction of the theory of the social logic of space with the space logic of society.

The need to understand the biunivocal socio-spatial relationship raises the following questions: is the success and consummation of the paradigm shift in the implementation of sustainable urban mobility mainly explained by the material essence (architecture—urban design) or the intangible (social sciences—change in people's cultural and social behaviour)? Or both, in the crystallization of a harmonious and unified communion?

In order to assess the most efficient typologies of urban acupuncture projects for the promotion of sustainable urban mobility, it is important to define an assessment process that frames the temporal value with the assessment criteria at the spatial, social, economic, and environmental levels. In this sense, the evaluation may consider aspects such as changes in individual and collective behaviour in the collective space, reinforcement of community spirit, functional transformations of the space or those at the level of materials, changes in security, urban image, and the perception of the city, and the contribution to decarbonisation, among others.

## 3.3 Implementation and conclusions

In view of the above, it is concluded that the concepts of time, mobility, and the humanization of the collective space are in tune with the reconfiguration of sustainable and resilient cities. In fact, the exemplary interventions at the level of worldwide good practices combine solutions that promote sustainable urban mobility, in the shortest time frame, and with repercussions at the level of training and collective involvement, with which, in turn, generate collective spaces that promote their urban experience.

On the other hand, it can be seen that, through urban acupuncture, the time of decline of a certain collective space is followed by a time of rejoicing and reference at the level of mobility and social cohesion of the same, with natural repercussions in other areas of the city.

The present approaches and themes were tested in 'Braga's Urban Innovation Laboratory', in four pilot projects (Mpt; CMB 2017) (figure 3.2): in the Montélios area (figures 3.3, 3.4, 3.5, and 3.6), Quinta da Fonte, the surroundings of Praça do Bocage (Torre Europa) (figure 3.9), and the Lamaçães urbanization (near Makro) (figure 3.8), financed by the Portugal 2020 Community Support Framework, within the development of PAMUS (Plano de Ação de Mobilidade Urbana Sustentável—Norte 2020—Sustainable Urban Mobility Action Plan—North 2020). These projects were carried out from 2017 to the present date and are currently at the end of the implementation phase and in the evaluation and monitoring phase of the processes to support replication in other areas of the city already claimed by the population. The methodological definition of the next interventions is being developed within a

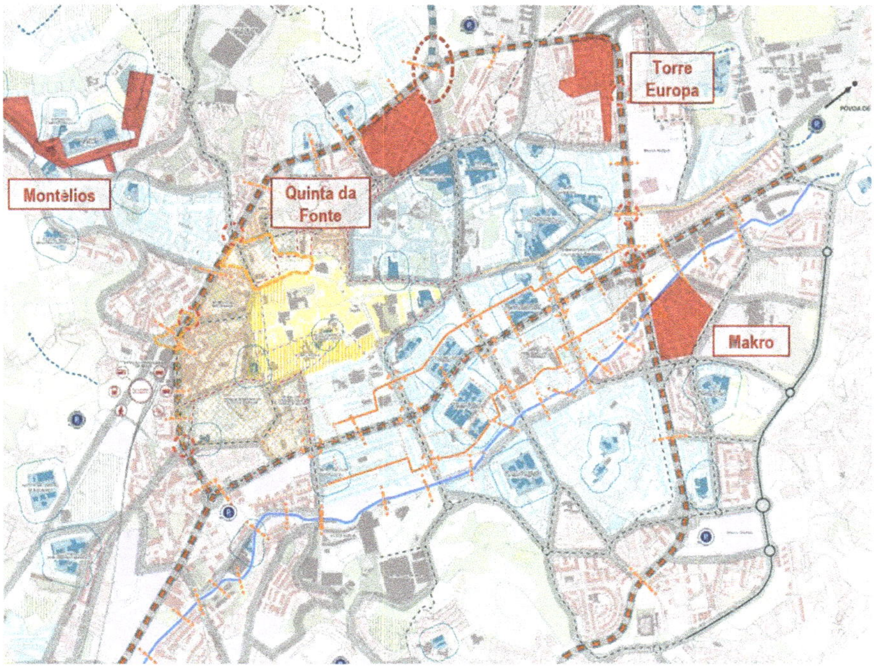


**Figure 3.2.** Planning map of the mobility strategy with the location of the four pilot projects (Montélios area, Quinta da Fonte, the area surrounding Torre Europa, and the area surrounding Makro). Source: Mpt; CMB (2017).

**Figure 3.3.** Aerial view of the Pilot Project intervention in the area of Montélios. Source: CMB (2019).

**Figure 3.4.** Aerial view of the Pilot Project intervention in the area of Montélios. Source: CMB (2019).

**Figure 3.5.** Moments from the public participation events held in September 2018 in the public space (source: the authors).

plan of cooperation between the mobility technical services of the City Council of Braga and the School of Architecture of the University of Minho, and with the additional participation of technicians from other cities that are part of the networks of European cities (URBACT and BEACON), of which the Municipality of Braga is a member (figure 3.5).

Besides the specific urbanistic process, the conciliation of the three concepts that are the basis of the interventions made are embodied in the processes consisting of temporal and/or periodic events organized in the collective space, which aim to promote sustainable urban mobility. These took place during European Mobility Week and used public participation processes. Some of these events created

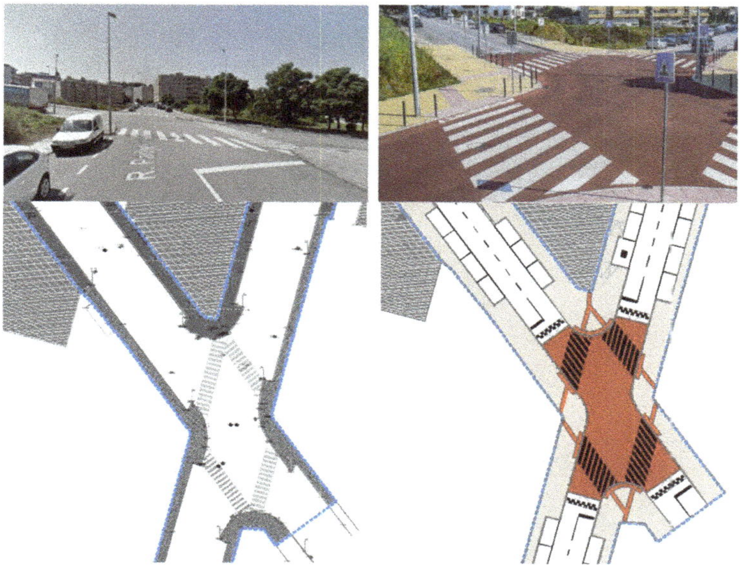

**Figure 3.6.** Before and after—pilot project of Montélios. Source: Mpt; CMB (2017); https://www.google.pt/maps.

**Figure 3.7.** Before and after—pilot project of Montélios. Source: Mpt; CMB (2017); https://www.google.pt/maps.

behavioral continuities, while others aimed to creating awareness of certain groups in society. However, the reach of these single-point events in time and space is materialized in structural changes in society and collective spaces throughout time.

The survey carried out among the population, already in a pandemic context, shows the receptivity of the respondents to switch to a more sustainable mode of travel (about 67%), and, without underestimating the possible contradiction between

**Figure 3.7.** (Continued.)

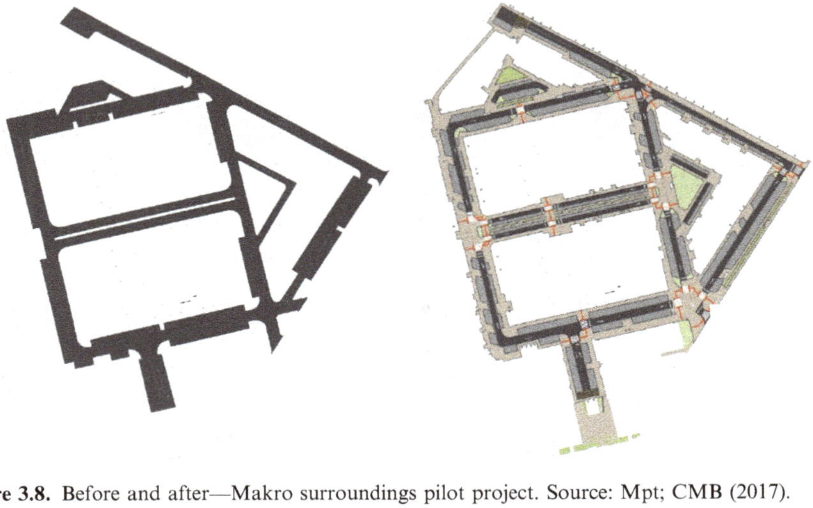

**Figure 3.8.** Before and after—Makro surroundings pilot project. Source: Mpt; CMB (2017).

the desire expressed by the survey participants and the actual execution of that desire, the result and the openness shown are an added value and a stimulus for municipalities to implement the inversion of the mobility pyramid with effective reflections on the modal split. For this purpose, it is necessary to undertake an active transition strategy with effective stimulus and dissuasive measures that include bold political measures properly contractualised with the different representatives and associations, in order to involve citizens in the decision-making and empowerment processes.

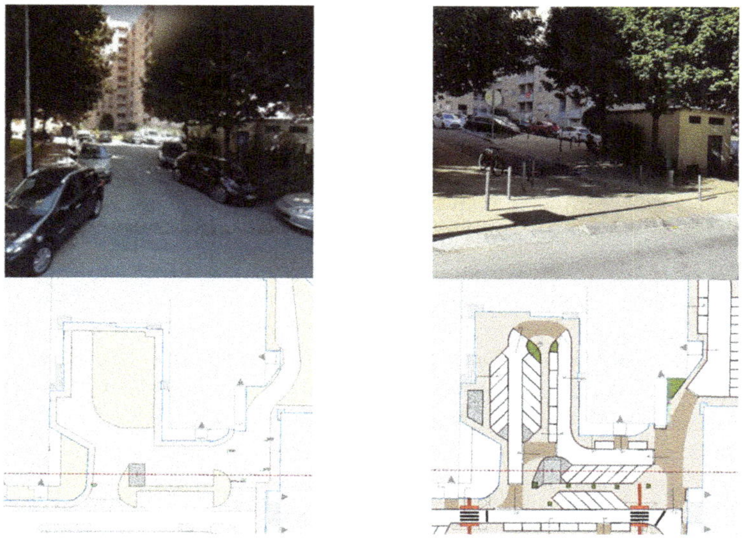

**Figure 3.9.** Before and after—Torre Europa surroundings pilot project. Source: Mpt; CMB (2017); https://www.google.pt/maps.

In short, it can be seen that the new planning model, anchored in actions of urban acupuncture and of increasing social cohesion, provides new characteristics to time. On the one hand, the interventions are quicker to implement compared to those framed in traditional planning processes; on the other hand, it can be seen that the communities now devote more time to using the collective spaces resulting from these methods. This is due to the increased identification of the population with the solution, through the process of participation and their involvement. This social rooting in urban design reflects a timeless character to the interventions. In fact, in contrast to the flexible, experimental, and low-cost character that is often inherent to this type of intervention, its effectiveness reveals a temporal perenniality.

Thus, challenging the radicality of the concepts of mobility in the face of notions of past and contemporary Time, the integration of the various agents in the decision leads to the crystallisation of the time of change in a near future.

Testing continues its way.

## Bibliography

Ascher F 1998 *Metapolis: acerca do futuro da cidade* (Oeiras, Portugal: Celta Editora) translated from French to Portuguese by Álvaro Domingos

Bornat D and Shaw B 2019 *Neighbourhood Design. Working with Children towards a Child Friendly City* ZCD Architects, London https://www.zcdarchitects.co.uk/s/Neighbourhood-Design-7p6d.pdf

Cervero R and Kockelman K 1997 Travel demand and the 3Ds: Density, diversity and design *Transportation Research D: Transport and Environment* **2** 199–219

Corais F 2016 Leituras Morfotipológicas e os IGT's *Conferência da rede Lusófona de Morfologia Urbana, PNUM 2016 (Centro Cultural Vila Flor, Guimarães, Julho 15-16, 2016)* 327–43

Corais F 2021 *A Cidade a 'Caminhar' para 2050. Braga como Laboratório para um Sistema Urbano Resiliente e Sustentável (Projeto Tese).* (Universidade do Minho)
CMB 2019 Brochure presenting the 'Área Mais' of Montélios, Braga https://www.facebook.com/municipiodebraga/posts/3506258052738690/
de Solà-Morales M 2008 *De cosas urbanas* (Barcelona: Editorial Gustavo Gili)
Gehl J 2011 *Life Between Buildings: Using Public Space* (Washington, DC: Island Press)
Gehl J and Svarre B 2013 *How to Study Public Life* (Washington, DC: Island Press)
Hillier B and Hanson J 2003 *The Social Logic of Space* (Cambridge: Cambridge University Press)
Honey-Rosés J *et al* 2020 The impact of COVID-19 on public space: an early review of the emerging questions—design, perceptions and inequities *Cities & Health* **Special issue** 1–17
Lerner J 2003 *Acupuntura Urbana, Investigaciones geográficas* (Rio de Janeiro: ABDR)
Mpt; CMB (Câmara Municipal de Braga) 2017 Projeto 'Área mais'—Zona Montélios
Pordata 2018 https://www.pordata.pt/Portugal/Ve%C3%ADculos+rodovi%C3%A1rios+motorizados+em+circula%C3%A7%C3%A3o+total+e+por+tipo+de+ve%C3%ADculos-310
Roorda C, Wittmayer J, Henneman P, Steenbergen F V, Frantzeskaki N and Loobach D 2016 *Transition Management in the Urban Context (Guidance Manual)* Drift (Dutch Research Institute for Transitions, Erasmus University Rotterdam)
Sengers F, Wieczorek A J and Raven R 2019 Experimenting for sustainability transitions: a systematic literature review *Technological Forecasting and Social Change* **145** 153–64
Silva C F 2014 *See(d)(k)ing Time: An Approach to How to Design as Research* University of Minho http://hdl.handle.net/1822/35034
Valentino-Devries J, Lu D and Dance G 2020 Location data says it all: staying at home during Coronavirus is a luxury. The New York Times 1–16 https://nytimes.com/interactive/2020/04/03/us/coronavirus-stay-home-rich-poor.html

**IOP** Publishing

Sustainable Urban Development

Topics, trends and solutions

**Luís Bragança, Cristina Engel de Alvarez and Luisa F Cabeza**

# Chapter 4

# New trends in urban mobility

**Begoña Guirao and Daniel Orellana**

This chapter reviews the trends in urban mobility over the past decade in Latin American and European cities. New Information and Communications Technology (ICT), advances in artificial intelligence, innovative business models, micromobility, and a renewed interest in active mobility (AM) have shaped these trends. For example, ICT that has been applied to public transport operation and management has decisively affected demand. Smartphones allow users to obtain real-time information about routes and schedules in large cities. The automotive industry is adopting less-polluting manufacturing processes, enhancing the offer of electrically powered vehicles, and improving technology for more autonomous and safer cars. On the other hand, active mobility, specifically cycling and walking, has experienced a renewed interest in cities over the past decade. AM has a minimal environmental impact, presents the highest cost–benefit balance, has demonstrable positive effects on the physical and mental health of the population, and is socially fair. The COVID-19 pandemic has affected these trends. On the one hand, it has accelerated the promotion and adoption of AM, and on the other hand, it has strongly diminished the use of public transport.

## 4.1 The role of public transport and active mobility in 21st century cities

The growth in travel by private car in intermediate and big cities since the end of the 20th century has raised serious concerns about air pollution, physical health, and public space quality. This growth led to the development of transport demand management (TDM) strategies to reduce private car use by promoting public transport as an alternative for users. The first TDM strategies focused on push measures to reduce private cars (i.e. tax increases on fossil fuel, congestion pricing, parking pricing, or vehicle use restrictions). However, many academics (Marshall and Banister 2000) soon started to emphasise the need to combine push measures with pull measures, mainly focused on public transport (i.e. public transport

doi:10.1088/978-0-7503-3971-1ch4  

improvements and ticket price reductions, among others). Some years later, Litman (2003) elaborated a TDM Encyclopedia, which was a study of the types of implemented measures that reduce private car usage, and listed 48 different strategies already implemented. Some strategies included measures that focused on public transit, active mobility (such as traffic calming, cycling infrastructure, or bike–transit integration), carsharing, and ridesharing. During the last two decades, these latter strategies have regained importance as political pressure to reduce pollution in cities has increased. More than half of the global **urban population is exposed to air pollution** levels at least 2.5 times higher than those defined by safety standards; transport pollution is a significant contributor to these figures. According to a United Nations database, figures in 2016 showed a devastating situation, with more than 91% of urban dwellers breathing unsafe air, resulting in 4.2 million deaths reported to be related to air pollution. However, air pollution is not the only threat related to motorised mobility. The 2030 Agenda for Sustainable Development adopted by the United Nations in 2015 stated the 11th Goal: 'making cities and human settlements inclusive, safe, resilient and sustainable'. Therefore, apart from being sustainable and clean, travel options should also be inclusive, safe, and resilient.

The technological change that the urban transport environment is experiencing, together with the emergence of the collaborative economy and the high penetration rates of smartphones among transport users seems to create an optimistic scenario in which to develop inclusive, safe, resilient, and sustainable transport. However, many of these changes (such as the development of autonomous vehicles or electric autonomous vehicles) are an ongoing process and are still subject to research and investigation. Therefore, it is too early to forecast future adoption behaviour or predict how these changes could affect the whole transport system, including public transport.

The role of public transport, which has been a cornerstone of ensuring equity and reducing pollution since the end of the last century, may change in this new scenario. The holistic value of public transport is often underestimated, as statements regarding its value are often focused on pollution reduction. Apart from its ability to transport large numbers of passengers efficiently and to reduce pollution, an efficient public transport system reduces other negative impacts of transport in cities while guaranteeing global accessibility and social equity. The potential societal benefits of improving public transport (mainly comfort and reliability) usually create an increase in demand, and this increase is usually underestimated by traditional studies, in which these quality indicators are dismissed.

The COVID-19 pandemic has shown the fragility of public transportation systems worldwide and has caused an unprecedented decline in demand (Tirachini and Cats 2020). The need to maintain physical distancing has challenged the space efficiency of mass public transportation, creating a perception of public transportation as unhealthy. Research agendas, policy guidelines, and practical recommendations are quickly emerging to support public transport recovery (Tirachini and Cats 2020). Simultaneously, active mobility has regained popularity, as shown by an unprecedented increase in cycling rates and walking in several cities across the world. Active mobility represents a safe and affordable alternative for large segments of the population, and several cities have implemented or accelerated their plans for new cycle lanes and enhanced pedestrian streets. Moreover, these kinds of infrastructure have positively affected the quality of public space and the population's physical health. Whereas walking and cycling may represent a feasible solution for short-distance mobility, longer distances still require motorised mobility. Current approaches advocate multimodality by combining active and public transportation, and some authors have underlined that this is the best combination to meet physical activity recommendations.

There are essential differences between the current trends of mobility in European and Latin American cities. Whereas European cities have relatively good public transport standards, Latin American cities (with some important exceptions) usually present lower levels of transit service. At the same time, active mobility in Latin America is an everyday reality for a significant proportion of the population, mainly due to economic conditions, while in European cities, the trend of active mobility is related to health and environmental motivations. ICT also represents an important difference between the cities in the two regions, with European cities embracing and rapidly adopting several innovations under the umbrella of 'smart mobility', including autonomous vehicles, the Internet of Things (IoT), and technology-based shared vehicle systems. On the other hand, most Latin American cities still lack robust information systems to support efficient planning and the expansion of public transport, and the incipient adoption of technology-based shared vehicle systems is confined to a small proportion of the population. Still, several cities are focusing on improving their mobility systems by increasing and organising public transport's fleet capacity, implementing Bus Rapid Transit systems, and enhancing cycling infrastructure. Despite the essential particularities of the cities in both regions, there are several commonalities between them, and many approaches might be applicable for both, especially for intermediate and big cities.

## 4.2 Disruptive changes: new vehicles and innovations in information and communications technology

The automotive industry is now facing significant changes created by the development of electrical powertrains and autonomous vehicle technologies, in the hope of producing more ecological, safe, and affordable vehicles. Although research into electric and autonomous vehicles started many decades ago, these technologies have now reached a minimum development level which makes their widespread use feasible. The **implementation of electric cars** still faces two main problems: the high market price of the vehicle (unaffordable for most households) and the limited number of charging points at urban and interurban levels. Electric cars are perceived as a new type of transportation mode characterised by their 'limited range' (associated with the distance to recharging points), and owners are experiencing so-called 'range anxiety'. These facts suggest that a long transition time will be required in order to reach a fossil-fuel-free transportation scenario in urban environments.

The automotive industry has reached a remarkably high level of innovation in the field of **autonomous vehicles**; nevertheless, the 'driverless' model expected by transport planners is still far away. The Society of Automotive Engineers has defined six automation levels (from 0 to 5), and this classification has been adopted worldwide. These levels are distinguished by the onboard driver assistance systems, i.e. the distribution of driving tasks between the driver and the vehicle. In vehicles with automation levels 0 to 2, the driver is still in charge of monitoring the environment. Level 3 is the threshold from which the vehicle is in charge of interpreting all the necessary data collected from the environment. Moreover, at this automation level, the vehicle can assume the driving task within certain limits. The highest level of automation is 5, in which vehicles perform the whole driving task autonomously, on all types of road, at all speed ranges, and under any weather conditions. Apart from prototypes, even level 3 is still far away. As in the case of electric vehicles, an extended transition period is to be expected before we reach widespread implementation of automation level 4. Future autonomous vehicles will be electric, in order to maximise all advantages provided by both technologies: fewer accidents, reduced congestion, and less pollution. In the autonomous vehicle scenario, each vehicle should interact not only with the road infrastructure (road layout, signalling, etc.) but also with the other vehicles and the rest of the environment.

New ICT has enabled the fast development of autonomous cars during the last decades, as it allows better data exchange between vehicles and between a vehicle and the transport infrastructure. ICT, together with the high penetration of smartphones among the population, has also favoured the emergence of new mobility forms based on the **sharing economy**, for example, vehicle sharing without a driver (carsharing, bike-sharing, moto-sharing) or with a driver (such as Uber and Cabify as forms of ride-hailing). Traditionally, ICT enhancements were first implemented in public transport and included fleet management systems based on automatic vehicle location technologies. They were also used to optimise routing and scheduling and to feed real-time information into passenger information channels. However, in the last decade, backed by the emergence of the sharing economy, the car rental market has also begun to benefit from new ICT. Smartphone apps enable the user to search for and book a car (or a bicycle, motorbike, or scooter) close to their current geographic location inside a geofence (a virtual perimeter) defined by the operator. Flexibility is, therefore, the main attraction and the differentiator of these free-floating carsharing (FFCS) systems. Nevertheless, there are still many unanswered questions about whether these systems increase the overall vehicle miles travelled by allowing easier one-way travel (automotive commuting) in urban areas, while negatively affecting traditional public transportation. The FFCS fleets can be totally electric (Amsterdam and Madrid), totally fossil-fuel powered, or hybrid (Berlin, Stockholm, Vienna, and Munich).

Furthermore, the automotive industry has also been responsible for the **emergence of these new forms of carsharing mobility**. So far, two major operators have dominated the free-floating carsharing market in Europe, North America, and Asia: car2go and DriveNow/ReachNow; they belong to car manufacturing firms (Daimler and BMW). One of their goals is to initiate customers into driving electric vehicles. Another goal for the automotive industry is to be a provider of this type of 'product service system (PSS)' in transportation and mobility. PSS growth has mainly been motivated by the need for traditional manufacturing firms (in this case, car manufacturers) to cope with changing market forces and to recognise that services combined with products could provide higher profits than products alone. A significant number of functional relations are changing within the automotive sector; the new strategy pursued by car rental firms is only one example. For many years, car rental firms were major customers of carmakers in the interurban scenario, but both are now competing for the same market.

All these disruptive changes (electric and autonomous vehicles, and new forms of mobility) are taking place simultaneously and represent a difficult challenge for the analysis of future scenarios for planners and policymakers. Furthermore, the idea of a future universal system of autonomous, electric, and shared cars is often presented as a 'panacea' that will solve all urban mobility problems, underestimating or dismissing public transport's key role in future scenarios. Many academics have started criticising these views, warning about potential hazards to congestion, social equity, and urban sprawl generated by a widespread system of electric, autonomous, and shared cars, and claiming that public transport should remain at the centre of urban mobility (Currie 2018).

New ICTs have impacted mobility, not only in its operational aspects, but also in terms of the methods used to analyse and manage mobility. The next chapter is devoted to describing the changes that new ICTs have brought and how these alternative approaches present scientific challenges.

## 4.3 The influence of ICT on mobility analysis and management

Human mobility research has gained new zest in the last decade, fostered by the synergy of new nontraditional data sources and new analytical methods based on geoinformation and the data sciences. A plethora of collection techniques, data models, analysis algorithms, and visualisation techniques have emerged to retrieve and analyse the ever-increasing amounts of geolocation data produced by mobile phones, web applications and services, social networks, and smart devices. These techniques complement the more traditional methods used by the social sciences and transportation engineering, which usually dominated this area of study until recently: self-applied surveys and 'travel diaries', and on the other hand, manual and automated gauging. These traditional methods, although necessary, involved significant logistical efforts, and researchers have questioned their precision for several years.

The immense **popularity of geo-enabled mobile devices**, social networks that incorporate location-based services (LBSs), and the possibilities of crowdsourcing and voluntary geoinformation (VGI) represent a whole new range of potential sources of mobility data. In the first place, global positioning technologies, such as the universally known Global Positioning System (GPS) and the modern Galileo system, are useful for recording the trajectory of one or more people simultaneously over extended periods in almost any part of the world. Applications such as STRAVA, WAZE, and KAPPO allow users of GPS-enabled devices to contribute to aggregated trajectory databases which are used by the app owners to provide LBSs and to market aggregated datasets.

Other positioning technologies, such as **network-based location**, have become prevalent in mobile devices, such as smartphones, tablets, and smartwatches. Companies such as Google, Apple, and Facebook actively collect their users' positioning data (sometimes without their complete awareness) and aggregate them to analyse individual and collective behaviour with the declared intention of providing targeted advertising. Additional data derived from location-based services can enhance the understanding of mobility patterns. Users of applications such as Twitter, Instagram, FourSquare, or Panoramio actively and openly share information about their daily lives, often adding data about their spatiotemporal locations. These data can be retrieved through programming interfaces and analysed to obtain mobility data, such as origin–destination matrices or urban space uses. These aggregated and anonymised datasets have proved extremely useful in the study of per-population mobility behaviours, such as domestic travel, urban traffic, or cycling patterns and, more recently, to investigate the changes of mobility during the COVID-19 pandemic. Another promising source of data is the automated analysis of video captured by security cameras in public spaces using computer vision algorithms. Several investigations have shown the possibilities of this technique to study interactions between cyclists and motorised vehicles or to extract information about pedestrian traffic.

Although promising, these sources of information have certain limitations. The first, and probably the most worrying, is related to **privacy**, since the increasing amount of data at the individual and aggregate level allows the derivation of personal information about users, an issue on which several researchers have focused (Giannotti and Pedreschi 2008). For example, several studies have shown that it is possible to reconstruct the activities, events, and paths of web service users via spatial analysis techniques and data mining. Moreover, the combination of this information with facial recognition technology allows a person's identity to be inferred from a photograph in real time. The loss of privacy due to the use of mobile devices has been a subject of public discussion for several years. One of the best-known cases is that of Malte Spitz in Germany, who sued his telephone company to obtain a copy of the records that the company kept on his mobile phone activity. The newspaper *Zeit Online* analysed these data and demonstrated that it was possible to obtain a detailed profile of Spitz's private and public life, including his home address, the streets where he liked to walk, his favourite bars, and his free-time activities. The compilation edited by Giannotti and Pedreschi (2008) presents an in-depth analysis of the relationship between the mining of mobility data and privacy.

Another critical issue with technology-based datasets and methods is the **potential socioeconomic bias**. For example, mobile-phone-derived datasets hardly constitute a representative sample of a city or country's population, due to the differences in technology adoption in different population segments. Despite the near-universal prevalence of mobile phones, the data they produce largely depends on connection events. Therefore, high-income population segments tend to present higher rates of online time and connectivity, and consequently, they are overrepresented in these kinds of dataset. Nevertheless, studies suggest that information derived from mobile devices can be robust for population density estimates and for identifying and extracting mobility patterns. Recognising and investigating these limitations is crucial in countries where technology adoption is uneven across populations.

## 4.4 Motorised mobility: trends and discussion

Future scenarios for motorised mobility show the predominance of shared, autonomous, and clean (electric or hydrogen fuel cell) vehicles over the other operational options and usually forget the role of public transport. Many authors (Currie 2018; Enoch 2015) have already raised concerns related to the feasibility of these proposals if the transport system is reduced to a universal system of shared electric 'automated taxis' (robo-taxis or taxibots), operated by mobility providers. Automated taxis will replace private cars (due to externalities) and conventional taxis (due to their lower cost), but they will also replace public transport because buses and trains will fail to satisfy point-to-point demand. In these scenarios, several problems arise from some characteristics that are not usually linked to autonomous, shared, and clean vehicles: the limited capacity of the system, the unavoidable congestion, the generation of a high number of non-shared trips, and the long-term, urban sprawl. In this dreadful scenario, to avoid road congestion, local administrations and operators might be forced to limit the number of trips per family, or to restrict the individual freedom to travel.

The **capacity of the transport system** is one of the main issues. There is no clear consensus that fleets of autonomous vehicles will improve the congestion and other traffic-related problems in cities. Although they are environmentally clean, autonomous electric vehicles still occupy space in streets, thereby affecting the level of service of roads or streets and the quality of public spaces. The existence of driverless cars will probably induce more trips. When exploring the future of autonomous vehicles (AV), research literature has recognised the need for urban rail systems that carry high volumes of passengers and cargo to complement any AV scenario and help cities to reach more sustainable mobility systems. Higher occupancy in vehicles is clearly a major advantage of public transport, and, as Currie (2018) pointed out, we can also work on the promotion of autonomous transit vehicles. In fact, AV train systems are nowadays fully operational for passenger services. The scenario of driverless shared cars should therefore be complemented by AV buses and trains.

The term **shared mobility** is widely linked to the idea of sharing trips, but in practice, shared mobility does not involve sharing trips. Ride-hailing (for example, Uber), bike-sharing, moto-sharing, and carsharing companies offer the option of sharing a vehicle, but sharing a trip is usually dismissed. When a vehicle is shared and the trip is also shared, the operation becomes highly efficient. But when a vehicle, even one that is 'clean' and shared, is only occupied by one individual, it still impacts traffic flow and congestion. Conversely, over 2000 people can travel on one train and 50 passengers can travel on a bus: this is real shared mobility. The concept of shared mobility that is efficient and environmentally clean should also be associated with public transport. Moreover, it is not clear that carsharing trips are replacing private car trips or public transport trips. The latter option will, in the long term, increase congestion and decrease the efficiency of public transport itself.

There is a common idea that shared mobility is growing and transforming shared vehicle use in cities. Nevertheless, the users of these services are still in a minority, and private cars still dominate the scenario. Moreover, the operations of carsharing companies are limited to some parts of a city (usually the central districts), and large segments of the population have no access to this kind of mobility service.

The quality of urban public space is also being affected by shared mobility. Free-floating carsharing systems require on-street parking in public spaces for an efficient operation. Bike-sharing and motorcycle-sharing schemes usually use the **sidewalk** for parking, generating many points of conflict with pedestrians and their safety. Sidewalks have now to cope with the presence of new obstacles and elements.

**Personal Mobility Vehicles (PMVs)** are also new actors in this future scenario of motorised mobility. The increased use of e-powered personal mobility vehicles (e.g., e-scooters, hoverboards, and monowheels) is popularly considered to be a positive change. Any electrically powered vehicle is readily accepted, as it is not supposed to have negative impacts on the environment. Nevertheless, some authors have sounded an alert that PMVs generate other kinds of problem, such as conflicts in shared public spaces, road safety issues, and traffic offences (Zagorskas and Burinskienė 2020). The increasing adoption of PMVs requires updated regulations and harmonisation strategies for shared spaces. Motorised mobility has, to some degree, invaded the sidewalk, which represents an entirely new and potentially dangerous situation for pedestrians. In the previous scenario, the sidewalk's curb was a clear frontier between motorised and non-motorised transport. However, when the frontier is unclear, road safety problems arise, mainly due to the differences in the speed and mass of the elements involved in a collision (pedestrians, bicycles, motorcycles, or scooters). Figure 4.1 shows the ranges of speed of different types of electric PMV together with their level of manoeuvrability (turning radius). The ranges of speed show that most of them (except the e-skateboard) can reach the speed of a car (45–50 km $h^{-1}$), compromising pedestrian safety and increasing the severity of any collision between a PMV and other motorised vehicles (cars or trucks).

While the idea of universal transport based on driverless, electric, and shared cars (solving all congestion and urban environmental problems) is reaching acceptance among populations, TDM strategies are focused on developing 'mobility as a service' (MaaS). An advanced MaaS can be understood as a flexible, integrated, and user-oriented mobility service developed through a digital platform, combining private and public transport providers. ICTs are, again, the base of this concept and a minimum level of digitalisation is needed among the population to achieve the objectives of MaaS. Sochor *et al* (2018) have produced a good analysis of this concept, looking for the more frequent keywords used to define MaaS: on-demand, user-centric, integrated, multimodal, flexible, platform, door-to-door, one-stop-shop, and seamless transport systems. But what is the actual level of implementation

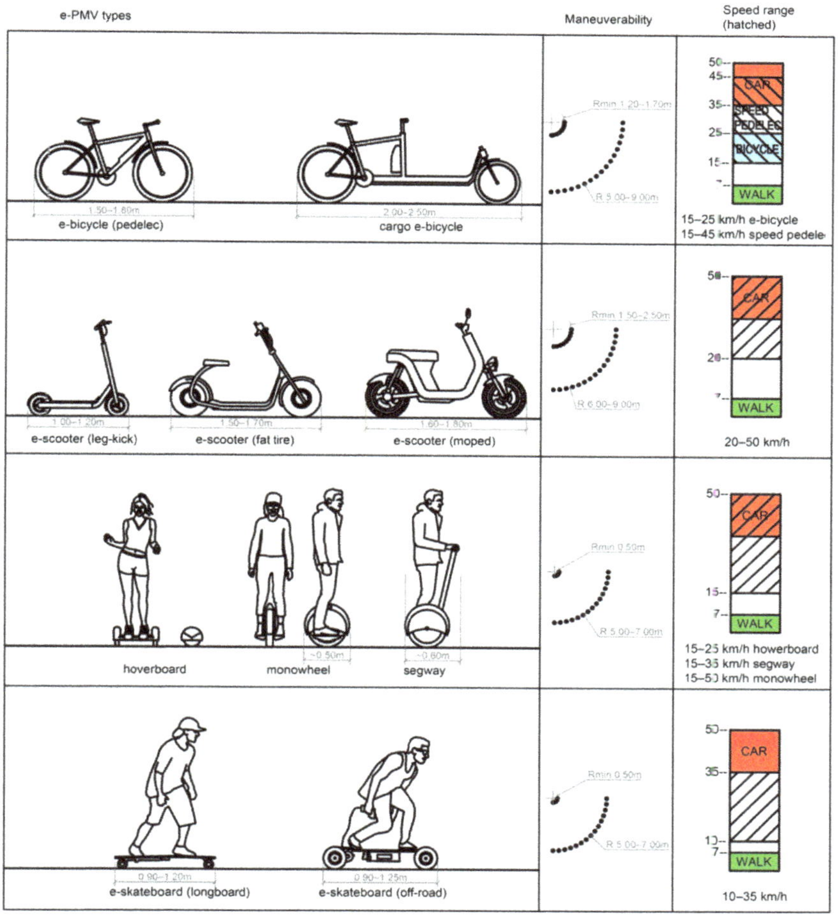


**Figure 4.1.** Operational speeds and manoeuvrability of the most popular electric PMVs. Source: Zagorskas and Burinskienė (2020). License: Creative Commons Attribution.

of MaaS in cities nowadays? A MaaS platform should help users not only to plan a trip, but also to buy a proper mobility package (a multimodal trip) covering the needs of different segments.

Nevertheless, in many cities, so-called MaaS are just platforms with which to plan trips and calculate travel times. Sochor *et al* (2018) identified four levels of 'integration' in MaaS. Level 0 implies no integration of services. In Level 1, the integration affects only information (multimodal travel planners, pricing information). In Level 2, only the integration of booking and payment is achieved (find, single trip, book, and pay). Level 3 can also achieve the integration of the service offer (subscription, contacts). Level 4 achieves the integration of societal goals, such as public policies and social incentives. Since Level 4 is rarely achieved, several authors are worried about the adverse effects that MaaS can generate if not organised correctly and if public administrations do not have a leading role in the organisation and management of MaaS platforms (Sochor *et al* 2018). Like the universal system of autonomous shared and electric cars, if MaaS is going to provide flexible travel, modal integration, and shared mobility, these new platforms can be expected to generate induced trips, compromising road capacity and congestion. Urban sprawl would consequently also be a long-term impact of MaaS implementation.

Given that the idea of MAAS is still far from being achieved in both Europe and Latin America, we can describe some differences in the mobility trends. As described before, European cities have relatively good public transport standards compared to Latin American cities, which usually present lower levels of transit service. Although transport apps have also emerged in Latin America, some kinds of shared transportation mode, for example, free-floating carsharing, are still difficult to implement, due to the low levels of transit service. Free-floating carsharing needs to be complemented by public transport to support a minimum level of demand. Furthermore, high car crime rates in some parts of the region also provide a disincentive for companies that plan on setting up carsharing schemes within Latin America. *Zazcar*, in Brazil, was the first carsharing system in South America, founded in 2009, and is still successfully in operation within Sao Paulo along with two other companies (*GlubGT* and *Vamos*). A few examples of carsharing can be found in México (*Carrot*) and Chile (*Awto*). In Europe, Germany, France, and Spain have experienced an emergence of FFCS systems in the last 10 years (together with moto-sharing and e-scooter sharing); with a growth trend, this is becoming one of the major FFCS markets in the world.

Unlike carsharing, ride-hailing has grown in Latin and South America, becoming a new commercial hotspot for ride-hailing companies such as *Uber*, *99 Taxis*, *Easy Taxi*, *Cabify*, *Beat*, and *Nekso*. The lack of efficient and accessible public transport systems, together with the security issues associated with traditional taxis has presented itself as the perfect opportunity for future ride-hailing business within this region. Uber has operated since 2013 in more than 16 Latin American countries; in Mexico, it is close to being the number one ride-hailing provider (Mexico is the second largest market after the US). *Maxi Mobility* is a company that owns *Cabify* and *Easy Taxi*, and operates across many Latin American markets, including Argentina, Mexico, Ecuador, Bolivia, Panama, Brazil, Peru, and Chile. In Brazil, *99 Taxis* (recently called *99*), backed by China's Didi Chuxing and Japan's SoftBank Group, has started operation and now has 14 million users.

Public transport and conventional taxis have also benefited from the emergence of transport apps and new ICTs. In Europe, efficient public transport has been reinforced by the provision of more real-time information to users (bus and metro arrivals and departures times at stops). In Latin and South America, the spread of transport apps has also been important in providing real- time information to users, although the provision of public infrastructure is still needed. Europe and Latin America share international mobility apps, such as Moovit. Moovit provides free information on the routes and time schedules of subways and buses in some 400 cities worldwide. Latin America currently accounts for 30 percent of the world's ten million Moovit users, and the company particularly sees Brazil, Colombia,

Argentina, Mexico, and Chile as strategic markets. The taxi market has also benefited from the implementation of these apps. For example, in Colombia, the app *Tappsi* connects users with yellow taxi drivers throughout the city, allowing one to order a cab (to be paid in cash or in-app) which can also be tracked using GPS for security purposes.

## 4.5 Trends in active mobility research

The prioritisation of active mobility, together with the improvement of public transport systems, have proved to be effective strategies for improving urban mobility. Specifically, the promotion of walking and cycling as a transportation mode helps to achieve the four main objectives of public space proposed by Gehl: vitality, safety, sustainability, and health (Gehl 2014). Simultaneously, **active mobility is economically efficient** for both cities and users, since the required investments in the acquisition and maintenance of infrastructure are orders of magnitude lower than those for motorised transport. Moreover, walking and cycling have meagre environmental and social costs, compared with travelling by motorised means, whose externalities have been estimated to account for around 18% of the average income in Latin American cities. On the other hand, **mobility by bicycle is socially equitable**, since it increases the autonomous mobility options for various social groups, such as minors and people with wheelchair mobility, who also benefit from cycling infrastructure.

Traditionally, the analysis and planning of urban mobility have been firmly based on transportation engineering approaches, and therefore have been conceived from an optimisation point of view. For example, it has been assumed that people move along the shortest routes in distance or time or those that are more efficient in terms of energy or cost. If this were the case, most people would travel on foot or by bicycle for short trips, and they would tend to choose the shortest route. However, **pedestrians and cyclists alike seem to follow patterns that differ from the optimal route**. Active mobility is more intricate than a logistical optimisation problem: is a complex behaviour that results from the interaction between individuals, groups, and their environment, and, as such, its analysis requires integrated frameworks and perspectives from different disciplines such as urbanism, psychology, engineering, ecology, and physical health, among others.

In this sense, like any other human activity, active urban mobility is influenced by many different factors that go beyond the time or distance invested in movement. During the last fifteen years, a large corpus of research has focused on these factors, which include culture, accessibility, safety, comfort, the pleasantness of the route, and urban characteristics, such as density, the mixture of uses, proximity to public transport and non-motorised mobility infrastructure, the topological and visual configuration of urban spaces, and the road network. Moreover, a person's mobility depends on their material circumstances, gender, age, ethnicity, abilities, and beliefs. So, the decisions of pedestrians or cyclists are closely related to their perception of themselves and their environment. In addition, the perception of the environment affects the selection of the type of mobility, routing behaviour, the choice of destination, and spatial segregation. Finally, planning, legal regulations, and institutional frameworks of cities, mainly respond to the needs of the private motorised vehicle, so the street has become a highly regulated and monitored space.

Alfonzo *et al* (2006) present an analogy between mobility and Maslow's pyramid of needs. According to these authors, accessibility constitutes the basic need that is to be satisfied using non-motorised mobility (base of the pyramid), understood in terms of connectivity, continuity, quality of infrastructure, and topography. In other words, accessibility can be a quantitative or qualitative variable that indicates how easily people can reach the places and locations that satisfy their needs or desires. Accessibility is followed in this pyramid by safety, comfort, and finally, pleasure (Alfonzo *et al* 2006); all these levels are interrelated. This analogy is used to organise a brief review of the relationship between the built space and cyclists' and pedestrians' behaviour and perception.

One of the main aspects of accessibility that influence the decision to walk or cycle is physical connectivity: a strongly connected street network tends to attract more active trips. Likewise, physical and visual continuity have an essential effect. For example, traffic lights and stop signs negatively influence cyclists' flow, because they increase travel time and energy expenditure. Research using spatial syntax theory and its tools has concluded that visual continuity and network interconnectivity are essential factors in the choice of a cycling route. Continuity is even more relevant for older adults, since the presence or absence of obstacles is decisive when choosing a walking route (Alfonzo *et al* 2006).

Likewise, the infrastructure's quality is a determining factor related to accessibility, one that influences the decision-making process about the choice of non-motorised mobility (Alfonzo *et al* 2006) and **infrastructure improvements reduce the impact of distance as a deterrent to cycling**. A similar thing happens with topography, since several investigations have demonstrated the importance of the gradient in the decision to walk or cycle. Although this is obvious for utility cycling, in sports cycling, the trend is the opposite, as slopes represent a motivation for training and fitness development purposes.

**Safety and security** emerge as the most critical characteristics of both utilitarian and recreational non-motorised trips. Deaths related to road accidents are considered to be a public health issue in several countries. Figure 4.2 depicts death rates from road accidents in various countries. Traffic-related deaths depend on lighting, intersection design, and interactions with traffic; the latter is of great importance, since speed directly influences the feeling of security of pedestrians (Alfonzo *et al* 2006). Whereas most cities in Europe but very few in the Americas have regulated motorised speed in urban environments, **traffic calming and speed reduction strategies** have been found to have a positive impact on the intensity and diversity of urban functions. They can improve the quality of streets and pedestrian environments, and function to spread out and organise pedestrian flows on wide streets (Rueda 2019).

It is also interesting to note that the construction of the perceptions of risk and safety is linked to group behaviour and age. Researchers have studied how the behaviour of others stimulates individual behaviour and found that the perception of

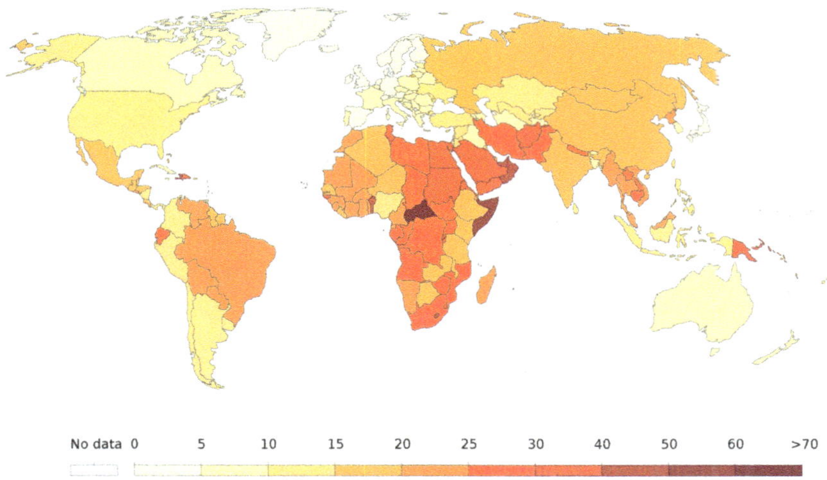


**Figure 4.2.** Death rate (age-standardized) from road accidents per 100,000 people, including those of drivers, passengers, motorcyclists, cyclists, and pedestrians, for 2017. Source: www.OurWorldInData.org. License: Creative Commons Attribution.

safety and the decision to take risky actions are determined mainly by the social information they have about the environment. On the other hand, it has been found that older people value horizontal and vertical signage much more, since this helps them to perceive an enhanced sense of safety. In contrast, young people take more risks, even given the possibility of making mistakes or breaking the law.

The decision to make a trip by bicycle is directly related to both objective safety, measured in terms of the number of accidents, and subjective safety, which is related to the perception of infrastructure safety. Thus, segregated cycle lanes are remarkably mostly preferred by women. According to some authors, however, total lane segregation is not always required. Physical segregation is suitable for trunk roads, but shared space is more appropriate for local roads. On a similar note, Gehl points out that '...in recent decades, ideas such as the integration and reorganisation of traffic have spread throughout the world. The latest innovation is shared streets, which work perfectly well as long as it is very clear that pedestrians have priority' (Gehl 2014, p 234). Despite the above, in the case of cyclists, some research indicates that although the safety factor is essential, it does not always guarantee an increase in bicycle use.

In addition to objective and subjective safety, comfort and aesthetics, which directly depend on urban design, are essential in non-motorised mobility. For example, visible squares, commerce, shops, and trees favour the decision to walk, while residential-only land uses discourage walking. Similarly, trunk and feeder roads have been regarded as environments that discourage pedestrian mobility. The urban landscape is what makes the difference between the use or non-use of a non-motorised environment. In other words, **the landscape strongly affects the decision about when and where to cycle or walk** but seems to have little or no effect on the use of motorised vehicles. The urban landscape is more than a background against which people perform their daily activities; it has agency and impacts individual and collective behaviour. It can be assumed that the images, sounds, smells, and sensations that come from the urban environment are perceived with greater intensity by pedestrians and cyclists, since the stimuli reach the senses directly, unlike the sensory capsule that a motorised vehicle represents.

The 'attractiveness' of the streets, defined as the capacity of a street or public space for social interaction and participation in cultural and commercial activities, is another factor that increases the use of non-motorised mobility (Gehl 2014). A street that can generate a point of attraction to stop, look, buy, and at the same time interact with other people, will indeed favour the use of non-motorised means (Gehl 2014). Hillier (2007) recognises the importance of 'standing people' as a crucial element of space culture, arguing that if people are standing, it is because something happened that made them stop: they found something that interested them, a site to rest, or they simply enjoyed a pleasant view. Just as silence is a fundamental part of music, stops are crucial elements of human movement.

Mehta (2008) incorporates the term 'utility' to refer to the significance of the diversity of proximate land uses for the individual. Studies suggest that the ability of an environment to satisfy day-to-day needs (shopping, food, or entertainment) increases the frequency of visits, making it more vital. For example, Miralles-Guasch (Miralles-Guasch 2011) reports that in Barcelona's central area, 89.9% of the population travel within the same neighbourhood to buy food, whereas this percentage decreases with the distance from the centre. Saelens *et al* (2003) explored the relationship between mobility and environmental and sociodemographic variables and concluded that the neighbourhoods' characteristics are relevant for different types of mobility. The authors found that inhabitants of communities with higher density, connectivity, and mixed use tend to use more active mobility.

Field research into active urban mobility has experienced accelerated expansion and diversification. Researchers have explored different factors affecting mobility behaviour, found and used new data sources, developed and adapted methodologies, and created integrated analytical frameworks. However, as the body of knowledge expands, so do new research questions; thus, it is expected that the new decade will be a sort of golden age for this field.

The implementation and promotion of active mobility have expanded along with the research field. Cities in Europe and more recently in Latin America are enhancing policies, plans, and infrastructure to promote walking and cycling. For example, Oslo, Copenhagen, and Brussels have implemented car-free areas where active mobility has replaced private cars, a program that is expected to reduce greenhouse gas emissions by between 60% and 95%. On the other side of the Atlantic, Bogotá has built 200 miles of bicycle lanes and closes an additional 75 miles of urban roads for active mobility once a week. An innovative approach to generating more walkable and cyclable streets was born in Barcelona under the name of 'superblocks'. Superblocks are groups of blocks in which the inner streets are pacified through urban interventions such as enhancing sidewalks, increasing vegetation, and restricting traffic speed to 20 km $h^{-1}$. This model has been shown to increase walking and cycling (Rueda 2019) and is being adopted by other cities such as Vitoria-Gasteiz (Spain) and Cuenca (Ecuador).

## 4.6 Conclusions and forthcoming challenges for 21st century cities

Urban mobility is undergoing an era of rapid change. After half a century in which the private car has been the iconic ideal of transportation in the city, there is overwhelming evidence of the various impacts of car-oriented mobility on the environment, health, and the quality of urban spaces, among others. As mobility is rising to become one of the greatest concerns among citizens, practitioners, researchers, and decision-makers, an era of innovation in urban mobility is growing across the globe, as evidenced by several changes that became common in the last decade: ICT is providing time- and location-aware information for users, while at the same time producing new rich datasets for research and planning; shared mobility became popular in most continents and challenged the established legal and social regulations for commercial transportation; private companies heavily invested in the development of autonomous vehicles with the questionable promise of solving urban traffic and improving safety. Alongside these innovations, cities must recognise public transportation and active modes as the cornerstones of efficient and sustainable urban mobility. New approaches are required to create synergies between technological innovations and longstanding solutions.

Policymakers should direct their actions on motorized shared mobility towards more efficient use of the whole transport system, using the available electric and autonomous driving technology. Public transport systems should also make use of autonomous driving and be non-polluting. Autonomous driving transit (bus, rail, etc.) can ultimately operate alongside and compete with autonomous shared-ride vehicles. Robo-taxis can be used clearly as replacements for transit on certain routes for service to disabled individuals, but the replacement of transit should not be the usual behaviour. TDM strategies should concentrate on allocating transit resources in corridors with more traffic congestion and with scarce on-street parking. Although autonomous driving promotes the use of motorized transport, attention should be paid to limiting the need for driving. Land-use policies and transport policies need to be coordinated through transit oriented development (TOD) strategies. Sidewalks and road crossings are also a matter of concern for policy-makers. Motorized mobility and active mobility can interact in the temporal and space dimensions, causing road accidents to result (fatalities and injuries). Moreover, the emergence of electric PMVs has generated a new type of accidental collision between conventional vehicles, pedestrians, and PMVs. Consequently, new road safety regulations may need to be investigated and implemented.

Active mobility can also benefit from innovations in technology and MaaS business models. For example, pedestrian-oriented neighbourhoods can improve the adoption of shared mobility as an alternative to privately owned vehicles, and algorithms for traffic optimisation can be adapted to prioritise cycling and pedestrian areas. Artificial intelligence algorithms can recommend walking or cycling routes based on historical movement patterns and personal preferences regarding aesthetics, safety, and security.

This chapter was drafted in the second half of 2020, six months after the lockdown caused by the COVID-19 pandemic. Mobility has not yet recovered to previous levels of demand, and working from home is still prevalent on a large scale across many countries. The authors have aimed to identify and document the long-term trends in urban mobility during the last decade, as a baseline with which to better understand the impacts of the pandemic. However, there are already several indications of the dramatic impact of the pandemic on mobility that will emerge in future years.

The authors believe that is a historic event that will probably affect mobility behaviour, not only in the coming months, but also for the next decade. On the one hand, massive public transport vehicles are perceived as environments with a high risk of COVID-19 transmission. Consequently, passenger use rates have plummeted during the pandemic and several countries have imposed limitations on the capacity of these vehicles. These factors have pushed public transport companies into financial difficulties, making it hard to maintain service levels. It is difficult to estimate for how long the impacts will last, but the situation represents a call to reimagine the way public transportation systems are designed and implemented. On the other hand, active mobility has dramatically increased during the pandemic, given the relatively low risk of transmission in open-air environments and the inherent physical distance required for cycling. It is expected that this increment will continue after the pandemic, since it represents an opportunity for cities to move towards more sustainable mobility. At the beginning of 2021, the impacts of the pandemic are still unrolling, and it is still unclear whether these changes will prevail as 'trends' or if they are temporary ad-hoc adaptations. It is still too soon to say. In any case, the pandemic has dramatically affected urban mobility, as it has affected other aspects of urban life.

## Bibliography


Alfonzo M, Boarnet M G and Day K *et al* 2006 *The Relationship of Neighborhood Built Environment Features and Walking* UC Berkeley: University of California Transportation Center http://escholarship.org/uc/item/8fh1x4h3

Currie G 2018 Lies, damned lies, AVs, shared mobility, and urban transit futures *J. Public Trans.* **21** 19–30

Enoch M P 2015 How a rapid modal convergence into a universal automated taxi service could be the future for local passenger transport *Technol. Anal. Strateg. Manage.* **27** 910–24

Gehl J 2014 *Ciudades para la gente* (Buenos Aires: Infinito)

Giannotti F and Pedreschi D 2008 *Mobility, Data Mining and Privacy: Geographic Knowledge Discovery* (Berlin: Springer Publishing Company) http://portal.acm.org/citation.cfm?id=1355342

Hillier B 2007 *Space Is the Machine* (London: Space Syntax Ltd) https://discovery.ucl.ac.uk/3881/1/SITM.pdf

Litman T 2003 The online TDM encyclopedia: mobility management information gateway. *Transp. Policy* **10** 245–9

Marshall S and Banister D 2000 Travel reduction strategies: intentions and outcomes *Transp. Res.* A **34** 321–38

Mehta V 2008 Walkable streets: pedestrian behavior, perceptions and attitudes *J. Urban.: Int. Res. Placemaking Urban Sustain.* **1** 217–45

Miralles-Guasch C 2011 Dinámicas metropolitanas y tiempos de la movilidad. La región metropolitana de Barcelona, como ejemplo *An. de Geogr. de la Univ. Complut.* **31** 125–45

Rueda S 2019 *Superblocks for the Design of New Cities and Renovation of Existing Ones: Barcelona's Case. Integrating Human Health into Urban and Transport Planning* (Cham: Springer Int. Publishing) pp 135–53

Saelens B E, Sallis J F and Frank L D 2003 Environmental correlates of walking and cycling: Findings from the transportation, urban design, and planning literatures *Ann. Behav. Med.* **25** 80–91

Sochor J, Arby H, Karlsson I and Sarasini S 2018 A topological approach to mobility as a service: a proposed tool for understanding requirements and effects, and for aiding the integration of societal goals *Res. Transp. Bus. Manag* **27** 3–14

Tirachini A and Cats O 2020 COVID-19 and public transportation: current assessment, prospects, and research needs *J. Public Transp.* **22** 1–21

Zagorskas J and Burinskienė M 2020 Challenges caused by increased use of e-powered personal mobility vehicles in European cities *Sustainability* **12** 273

**IOP** Publishing

Sustainable Urban Development
Topics, trends and solutions
**Luís Bragança, Cristina Engel de Alvarez and Luisa F Cabeza**

# Chapter 5

# Coexistence of urban regeneration policies and sustainable mobility: reflections on, and practical cases of, urban acupuncture

**Miguel Bandeira, Filipa Corais and Fátima Pereira**

The rise of concerns about the promotion of sustainable urban mobility, which is part of the political agenda at several levels and scales, from global to local, acts as a catalyser of urban regeneration policies. The socio-economic and cultural fields work alongside the physical and spatial ones to foster more sustainable behavior in the community. The approach was applied to a case study in Braga through successful 'urban acupuncture' interventions that, even though isolated and focused on single locations, because they promoted integration with the community, were also strategic and enabled chain reactions on a more global scale.

## 5.1 Introduction

The contemporary processes of the transformation of cities and, even more indiscriminately, of urban spaces in general, require a revisiting of places from the perspective of the recognition and consideration of pre-existing interventions. Rehabilitation is thus one of the main priorities of cities, regardless of whether it is applied to the buildings, the equipment, the infrastructure, or other more or less free spaces, more or less subject to social anthropism. This is imposed by the multilateral crisis of socio-economic expansionism, the ageing of our societies and, above all, the growing global awareness of environmental interdependence and its fractious and irreversible imbalances.

Look at the European context in which we find ourselves, first and foremost, in the context of the *urban community's collection*, where we can take, as a structural milestone, the Leipzig Charter (Bandeira 2020) on sustainable European cities, correlated with other benchmarks, such as the Declarations of: Marseille (2008) and Toledo (2010) on integrated regeneration; Riga (2015), with the promotion of a European urban agenda, which emphasises the key role of small and medium-sized cities; and finally, the adoption of the so-called Urban Agenda for the EU, with the Amsterdam pact (2016) (Bandeira 2020).

In Portugal, due to the convergence of policies with these goals, the Legal Regime of Urban Rehabilitation was approved (RJRU—Decree Law No. 307/2009 of 23 October), which provides municipalities with reinforced means for the pursuit of Urban Rehabilitation, through the creation of instruments and the approval of incentives and mechanisms for coordination and consultation between public and private entities/investments. According to this legislation, urban rehabilitation is assumed to be indispensable '*insofar as the objectives of requalification and revitalization of cities, in particular of their most degraded areas, and of qualification of the housing stock are converged in it, seeking a globally more harmonious and sustainable functioning of the cities, and the guarantee, for all, of dignified housing*' (RJRU—Decree Law No. 307/2009 of 23 October).

An area of urban rehabilitation is thus defined in the legislation as a '*territorially delimited area which, due to the insufficiency, degradation or obsolescence of buildings, infrastructures, equipment for collective use and urban and green spaces for collective use, namely, with regard to its conditions of use, strength, safety, aesthetics or salubriousness, justifying an integrated intervention, which may be delimited in a specific instrument or correspond to the area of intervention of a detailed plan of urban rehabilitation.*' It was under this understanding and due to the need for investment in urban regeneration to be included in 'areas of urban rehabilitation' (ARUs) that municipalities were encouraged to review and agree on a strategy in order to access the Community framework under the Strategic Plans for Urban Development and the Financial Instrument for Urban Rehabilitation. This represented an important step forward in terms of public policy, agreement on plans and instruments for urban regeneration, and intervention in disadvantaged communities and the field of mobility.

For the purpose of the RJRU '*the definition by the municipality of the objectives of urban rehabilitation of the delimited area and the appropriate means to pursue them is associated with the delimitation of the areas of intervention (the 'urban rehabilitation areas'). It is based on a broad concept of urban rehabilitation and special emphasis is placed not only on the real estate or heritage side of the rehabilitation, but also on the integration and coordination of the intervention, stressing the need to achieve coherent solutions between the functional, economic, social, cultural and environmental aspects of the areas to be rehabilitated. Hence, we begin by defining the essential objectives to be achieved through urban regeneration, and determining the principles to which it must conform*' (RJRU—Decree Law No. 307/2009 of 23 October), among which we highlight sustainable urban mobility. Therefore, the surroundings of the historic centre, lacking in public intervention, designed around the car and the 'mass' construction developed in the 1980s and 1990s, are the ideal targets with which to place the concepts of accessibility and sustainability on the agenda, which thus join the concern for the rehabilitation of the construction. And it is the public space which takes centre stage in the municipality's initiative.

Besides the building component, and the others, according to the legislation in question, urban rehabilitation, within the public space and mobility component, concentrates on the intervention, in an articulated manner, for the '*pursuit of the following objectives: To rehabilitate degraded or in degradation urban fabrics; To ensure equal opportunities of citizens in the access to infrastructures, equipments, services and urban functions; To requalify green spaces, urban spaces and equipments for collective use; To qualify and integrate especially vulnerable urban areas, stimulating social inclusion and territorial cohesion; To promote the general improvement of mobility, namely through better management of public roads and other circulation areas and to promote the creation and improvement of accessibility for citizens with conditioned mobility*' (RJRU—Decree Law No. 307/2009 of 23 October).

It is therefore within this framework that access to Community investment is offered, such as the Operational Programme for Technical Assistance (OPAT), co-financed by the Fundo Europeu de Desenvolvimento Regional (FEDER), in other words, one of the support instruments for programmes in Portugal 2020, provided for under the Strategic Urban Development Plans (PEDU), which have required investment in mobility and urban regeneration to be made compatible through the obligation to elaborate three types of 'plan': a sustainable urban mobility action plan (PAMUS); an action plan for urban regeneration (PARU); and an integrated action plan for disadvantaged communities (PAICD). In Portugal, this cooperation enabled projects to be structured into these three complementary areas, where the objectives of mobility and urban rehabilitation were duly contemplated.

## 5.2 The benchmark for sustainable urban mobility

The global awareness of climate change, and, in particular, the effects of the rise in temperature and the relevance of its connection with decarbonisation, is claimed to be a consensual necessity and is imposed independently of the commitments made by the government for 2050. Indeed, cities are in a process of population growth, with resulting consequences for accessibility (Banister 2011; Bertolini 2020), mobility, and quality of life in general, which require the adoption of sustainable, albeit non 'canonical', solutions (Corais 2021, p 16).

For this reason, we consider the role of architecture in the interdisciplinary component of mobility policies, which have been neglected, both in the theoretical and the practical field; it is therefore urgent to assume it as a structural dimension. The link between architecture and mobility, complemented by other inseparable disciplines, is based on the principles of spatial organisation, especially in terms of the collective space system, design, and the urban morphology of cities (Corais 2021, p 16).

de Solà-Morales (2008, p 148) asserts the importance given to urbanity by diversity and density, while valuing, in the latter, the qualitative attributes (referring to the variety of overlaps in each location) in relation to the so-called quantitative. The same sense as that of the systematisation of 3D (Diversity, Density and Design), the theory presented by Cervero and Kockelman (1997), corroborates the thought that defends the influence of an environment based on commuter travel. Hence, urbanism begins to integrate mobility as a substance of architectural design, as opposed to the concept of the segregated zoning road network (de Solà-Morales 2008). This theory of the symbiosis between architecture and mobility is therefore later tested and synthesised by Manuel de Solà-Morales (2008), who attributed to architecture the role of controlling the impact of flows in public spaces which stimulate social practices (Corais 2021, p 18).

## 5.3 And the benchmark of urban regeneration

From the same synthetic perspective, it is important here to provide a sketch of Urban Regeneration, in particular, its convergence with sustainable urban mobility projects, and to pool the energies of both urban policies in the system of collective spaces.

Referring once again to de Solà-Morales (2008), by means of the theory of urbanism, this author approached the assumption of the dichotomy between public and private areas, affirming the prevalence of the concept of collective space in place of the former. According to this author, the supremacy of the public versus the private area, which emerged at the end of the 19th century and has continued until the present, is linked to urban-political ideologies that have taken shape and have integrated various theoretical contents over time, but are nowadays unsuited to explain the reality we live in. In this evolution, there is a false focus of the consolidation of the city's form, centred on dominance, when the core lies in functionality, that is, the connection of the spatial logics of the urban system itself. In this regard, de Solà-Morales (2008) attributes the vital importance of city areas to their urban experiences and to the social interaction that triggers and enhances the intrinsic appetite of private areas (such as public areas) for collective heritage. Paradoxically, this work also refers to the individual/private or collective appropriation of certain public areas in the city. Similarly, Gehl (2011, p 125), while placing emphasis on public areas throughout his speech, acknowledges the increasing trend towards collective use of private areas (e.g. commercial areas, shop arcades) (Corais 2021, pp 21 and 22).

The emphasis is therefore given to the collective spaces system that encompasses and overcomes the public space system. In fact, collective spaces, as an integrated system, assume an important function in the structuring and characterization of urban identity, as they structure the city formally and functionally (Corais 2016). As the system of collective spaces is the primary element of the urban form, it supports the city's model, enhances its spatial integration and manipulates mobility. However, urban regeneration also involves the rehabilitation of private buildings, which has had strong implications for sustainable urban mobility. Consider, for example, the problem of combining universal accessibility with the aesthetic factors arising from the preservation of the authenticity of the historical and cultural heritage, in terms of the shapes of structures, the impact of materials, colours, devices, and other collateral factors. At this level, it is worth highlighting the densification of buildings currently promoted by urban rehabilitation, which induces greater pressure from car traffic, particularly in the established areas of cities.

In this sense, besides urban regeneration, mobility also has as its scene of action the system of collective spaces that structures the city and in which the population actively interacts with its options and lifestyles.

Sustainable mobility is thus one of the main concerns when intervening in the public sphere today. The intervention model and the concerns and issues of mobility and rehabilitation depart from focusing on the car in order to focus on pedestrian and urban sustainability. Consider the numerous examples in Braga, and around the world, of converting parking areas and motorways into spaces for pedestrians and the community. This kind of intervention has gained momentum in the current pandemic situation, because of both the pertinent extra need for free and open space, which the new way of living demands, and the developing mobility patterns. Public space has started to rely far more on active mobility, aside from leisure functions, which includes commuting for both work and study (Bereitschaft and Scheller 2020). Cities' ability to be resilient has been put on probation through their response to the new social distancing challenges and mobility patterns, which entailed a need to (re)design public spaces in order for them to be able to accommodate the increasing flows.

In this respect, accessibility is crucial for the construction of social practices, for integration, and for equal access. '*According to the Constitution of the Portuguese Republic, it is the duty of the State to adopt measures whose purpose is to guarantee and ensure the rights of people with special needs, thereby promoting the well-being and quality of life of the population and real and legal-formal equality among all Portuguese [subparagraph d) of Articles* 9 *and* 13*], as well as to implement 'a national policy of prevention and of treatment, rehabilitation and integration of citizens with disabilities and support for their families' and 'to assume the burden of the effective realisation of their rights' (n° 2 of Article 71)'* (RJA—Decree Law No. 163/2006 of 8 August). The constitution secures the city the right to equality in the public domain.

In Braga, over the last decades, the urban growth model focused on the privilege of car circulation and 'mass' construction, based on the sum of isolated urban operations and site permits without being based on urbanisation plans and detailed plans. This model, which was the main one applied in the 1980s and 1990s, had decisive consequences for the design of public areas, and for many years, it compromised the right to universal accessibility established by law. In this context, it is essential to revisit these urban areas in order to achieve urban rehabilitation. This perspective is part of the principle of sustainable city policies, which consider energy efficiency measures, sustainable mobility, and accessibility for all, while contributing to the improvement of health, wellbeing, security, and population inclusivity.

## 5.4 Urban acupuncture applied to sustainable urban mobility and urban regeneration

The territorial approaches based on urban *acupuncture*, developed by Manuel de Solà-Morales (2008) and Jaime Lerner (2003), have very relevant application in the fields of mobility and urban regeneration, which deserves to be explored. According to de Sóla-Morales (2008), urban *acupuncture*, like medicinal acupuncture, acts strategically at neuralgic points and determines a much faster and more comprehensive capacity for action spread throughout the city (Corais 2021, p 29).

Urban *acupuncture* therefore refers to urban interventions that constitute a one-off stimulus in a given strategic location of the territory, which, on the one hand, present a problem to be solved, and, on the other hand, have the capacity/potential to create synergies within a broader urban context, primarily due to the catalytic effect of the transformation inherent to this very principle. As Daugėlaitė and Gražulevičiūtė-Vileniškė (2018) point out, energy flows applied to the urban fabric have various valences (e.g. social, economic, ecological, urban, cultural, among others). Or in the case of Hoogduyn (2014), there are eight defined principles that frame urban *acupuncture* actions: the definition of the *sensitive point* at which to act; the *scenario* or vision that frames it; the need for it to be a *quick act*, which includes the *participation* and involvement of the parties concerned; acting as mechanisms to promote *education*; and although the interventions are punctual, they have a *holistic and interdisciplinary approach*; being *small-scale* interventions, but tactical; and having as their purpose the *creation of places* (Corais 2021, pp 29 and 30).

Within this framework, it is hoped that the organisation of these one-off urban *acupuncture* interventions in the overall system of the city as a whole will find practicality in the adoption of sound principles of urban morphology and respect for the pre-existing forms of urban tissue, providing a motto for the promotion of sustainable urban mobility and urban regeneration.

According to Saverio Muratori (Corais 2016), the primary explanation for the viability of urban continuity and the organic evolution of the city is the structural concept of type in morphological approaches. In fact, the typomorphological approach as a tool for the promotion of sustainable urban mobility and urban regeneration, is, by its permanence, a guarantee of feasibility, greater ownership by the population, and the safeguarding of the city's identity.

Nevertheless, the short-term need to include the principles of sustainable urban mobility in new areas of expansion to ensure maximum effectiveness and efficiency of the paradigm shift means that it must be possible to implement them, especially in the established city. In fact, in order to achieve the desired modal shift from individual motorised transport to so-called soft modes and the regular use of public transport, in order to promote the decarbonisation of cities, the mobility strategy must provide for the integration and coherence of municipal policies in the framework of territorial cohesion. In this sense, mobility policies should be cross-sectoral and integrated in all interventions, especially in the most central areas of cities, and in those developments with a strong impact on the expansion frames that serve them, both by the catalytic effect they have in other areas and by mitigating the concentrated effects that occur in urban centres (Corais 2021, pp 25 and 80).

In this chapter, the sets of measures thus define infrastructure interventions (*Hard* measures), which improve the pedestrian and cycling conditions of cities (*Pull* measures), using the rescue (*Push* measures) for spaces that were affected by cars in the planning and design actions of the 1980s and 1990s, integrated into communication and awareness campaigns for these causes (*Soft* measures). The implementation of policies to promote sustainable urban mobility as part of a *Pull/Push/Hard/Soft* package of measures (Banister 2008) thus implies an equitable redistribution of public space between the various modes of transport (Corais 2021, pp 9, 30, and 31). This redesign of the urban form is a double opportunity for the city and for the quality of life of its residents, embodied jointly by urban regeneration and by promoting sustainable urban mobility.

The various mobility networks must then provide for their integration, connectivity, and legibility, as well as for articulation with the main mobility hubs. Urban areas, in general, present a concentration of equipment and services that are travel enhancers and generators of the most varied forms of travel. This confirms the above statement, i.e. to prioritise sustainable mobility interventions and to promote the integration of urban regeneration principles into such mobility projects. Whilst not neglecting the overall strategy defined for the territory in general, the promotion of sustainable urban mobility can be made operational through specific urban regeneration interventions, strategically located, and with a catalytic effect on the rest of the territory.

Hence, it is considered fundamental to associate urban morphology with mobility networks, and it is proposed that this articulation be made through the implementation of urban *acupuncture* projects (Lerner 2003; de Solà-Morales 2008; Hoogduyn 2014) integrated into a system of collective spaces in the city.

## 5.5 Amongst practical cases

Today, Braga is experiencing a development strategy based on the humanization of the city and the affirmation of concepts such as sustainability, in which decarbonisation, the improvement of air quality, in short, healthy living, and inclusion, among others, are ever-changing goals. To support these, in the various projects developed, an inversion of the mobility pyramid has been gradually applied, which consists of reducing the space dedicated to the car, and increasing the preponderance of pedestrians, highlighting soft modes and public transport in the urban design.

Braga has been built according to the register and the single allocation procedure (figures 5.1 and 5.2), which was progressively licensed under legislation highly encumbered with the privileged rights of private property. It is therefore in the surroundings of the historic centre, in the settled city, that we find the lower-quality neighbourhoods, a consequence of rapid growth and poor planning, where we perceive an urban environment in need of regeneration and where an integrated approach is required in terms of urban rehabilitation, mobility, and urbanisation. (In, https://ru.cm-braga.pt/0101/reabilitacao-urbana/aru-zona-consolidada/enquadramento.)

Urban rehabilitation, which '*until recently was confined to the physical preservation of the building and the public space of historic centers, and/or large heritage sites, social housing, now seems to defy new purposes and other scope. In addition to the construction, reconstruction, expansion, alteration, conservation or demolition of buildings, legal semantic spectrum, too imprecise but seems to include everything, also incorporating other types of consensus-based intangible assets in terms of quality of life, which are part of social integration, economic development, the tranquility of communities and the cosmopolitan diversity that enhances urban life. But also, in the case of Braga, it now focuses on new territorial fabrics previously excluded from the concept, such as the urban crown built in the previous century*' (Município de Braga; Quaternaire 2016.)

In fact, the central location of the ARU's defined for the city of Braga reinforce the peculiar character centered on this municipality in its dispersed territorial context, raising sustainable urban mobility to the status of a priority for urban rehabilitation policies and interventions in public spaces.

The 'I already pass here!' project (figure 5.3), financed by the Portugal 2020 Community Support Framework, within the development of PAMUS, focuses on

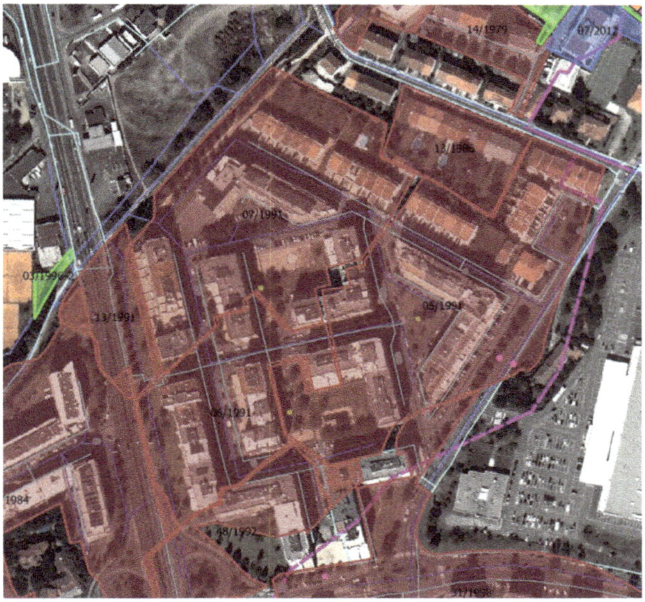

**Figure 5.1.** Overlapping allocation permits—Lamações area (Source: Municipality of Braga).

**Figure 5.2.** Overlapping allocation permits—Praça do Bocage area (Source: Municipality of Braga).

the consolidated network of the city. This is a bottom-up project, which began in the Participatory Budget (the proposal 'I'm not passing here!' alluded to the overcoming of architectural barriers that focused on one of the city's parishes), and which the municipality adopted and reversed under a positive emphasis, extending it to the

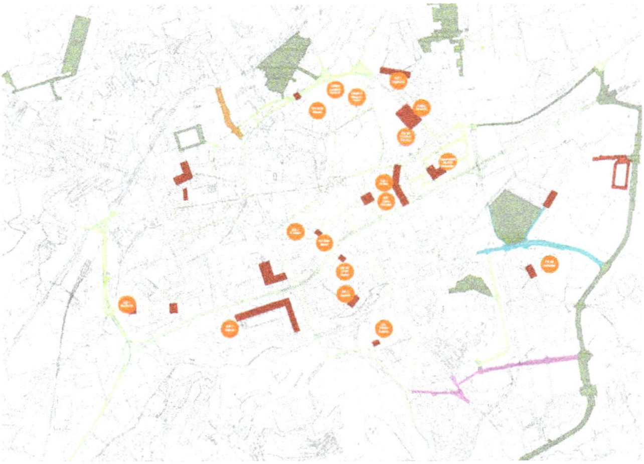

**Figure 5.3.** Location map of the interventions of the project 'I already pass here!' (Source: Municipality of Braga).

promotion of mobility and school safety, distributing the interventions in order to improve the connectivity of schools with the city center. This involved the surgical correction of about 80 pedestrian crossings, in about 19 specific locations, with the fundamental aim of promoting sustainable mobility, security, education, and inclusion.

Nevertheless, these interventions consist of topical actions, apparently disconnected from each other and framed in a global strategy of urban regeneration that aims at reversing the pyramid of mobility—that is, replacing the supremacy of the private car, at the top, to give priority to the pedestrian—and which is articulated with other ongoing projects and with the synergies that these will give to the collective space, guaranteeing the desired change in the pedestrian network.

At the same time, the municipality of Braga, committed to redesigning the urban environment of the established city, programmed the implementation of four pilot projects 'Areas +' (+ People; – Cars; + Quality of Life; – Noise) (Mpt; CMB 2017). The pilot projects included four areas in need of physical intervention: the Lamaçães area, Quinta da Fonte, Montélios, and Praça de Bocage.

The priorities for intervention in the pilot areas focused, in first instance, on the elimination of urban and architectural barriers, indicating the spaces with the potential for a change of use, namely new squares, playgrounds, and recreational or green areas. These projects started with public information sessions in the places themselves, with the residents, trying to improve the security and circulation conditions, and organising parking in public areas. In response to the expectations of the community in general, these areas of the city have become more inclusive, child-friendly, and sustainable. They have become places where families, the young, and the old can more effectively interact and give new meaning to the public space.

The changes proposed in this project will have a leverage role, as they will contribute to replicate the daily pedestrian habits throughout the city. In addition to the qualification of the pedestrian channels, various and complementary measures will be taken to promote the calming of traffic and contribute to increasing road safety on the city streets. Many of the convivial functions of the public space have been transferred both to the growing intimacy of the private domain, increasingly individualistic and oblivious to common problems, eroding the feeling of belonging to a place, a neighborhood, and the city. This is a fundamental condition for mobilizing the defeat of collective problems. The aim is to strengthen social ties in the neighbourhood and to experience the public space within neighbourhoods and urbanisations.

Jane Jacobs was a mentor in the 1960s for the awareness of these matters and, more recently, Jan Gehl (2011; Gehl and Svarre 2013) has been a faithful successor and diffuser in the world through the application of these theories in various cities. The humanized spaces of the city are capable of triggering urban experiences and social interaction, which consecrates them as a collective heritage.

Gehl and Svarre (2013, p 3), in line with the criticism undertaken by Jane Jacobs, in the 1960s, William H. White, in the 1980s, and Christopher Alexander of modern urbanism, advocated the humanisation of the city by reducing the scale of public spaces and reducing the presence of cars, favouring pedestrians and urban living; this took place against the backdrop of all the policy underlying urban design that sought to qualify an urban network designed for the car, speed, and interiorisation of the community within large housing blocks.

## 5.6 Concluding summary

The current state of the art of promoting sustainable urban mobility and the radical change that determines society's current travel habits and attitudes require us to consider the direction of this profound transition process. It shows a broad social, equitable, and intergenerational commitment leading to collective acceptance (of the priorities of the whole and not only of the individual) and the possible recovery of true community spirit.

Similarly, and at the same time, the prevalence of the implementation of these kinds of intervention in urban centres implies an articulation with the policies and principles of urban regeneration, especially with regard to the system of collective spaces, but also of private undertakings. This articulation is proposed to be surgical and incisive through urban *acupuncture* projects, which can have an exemplary effect, inducing new behaviors which are more far-reaching than the point at which they directly act. For this, these projects, although located at sensitive and specific points in the short term, are framed in a policy strategy of integrated action, in which the chain of actions and the replicating effect within the territory guarantee a cohesive and global intervention in the long term.

Due to urban dynamics, the spirit of unpredictability that defines this pandemic moment, and the economic impact inherent to typical actions of traditional planning that were structural in the territory, the use of urban acupuncture interventions is the most effective.

To maximize the effects of these interventions in the long term, it is necessary that they are integrated into a mobility planning strategy, that the definition of vulnerable points is careful and has an impact on the territory, and that there is an effective involvement of the community in the process, creating a sense of belonging and the necessary co-responsibility to ensure the maintenance and management of the places involved.

In short, the link that relates urban regeneration and sustainable urban mobility in the intervention projects that take place in the system of collective spaces, and which makes them inseparable, consists in the recovery of the local collective community's spirit and also in the contemporary field of action of policies based on complementary and decisive socio-cultural, economic, and urban transition commitments that are decisive for successful implementation, in all aspects.

## Bibliography


Bandeira M 2020 A carta de Leipzig: a cidade sustentável e a sustentabilidade da ideia treze anos depois *La Nueva carta de Leipzig. Una ocasión para repensar la política urbana de la Unión Europea* (Eixo Atlântico do Noroeste Peninsular) pp 63–97 https://coruna.gal/descarga/1453749920165/LaNueva-CartaLeipzig_AportacionesEixo_2020.pdf (in English: http://repositorium.sdum.uminho.pt/bitstream/1822/68387/2/MBENG.pdf) 978-989-54875-8-5

Banister D 2008 The sustainable mobility paradigm *Transp. Policy* **15** 73–80

Banister D 2011 Cities, mobility and climate change *J. Transp. Geogr.* **19** 1538–46

Bereitschaft B and Scheller D 2020 How might the Covid-19 pandemic affect 21st century urban design, planning, and development? *Urban Sci.* **4** 56

Bertolini L 2020 From 'streets for traffic' to 'streets for people': can street experiments transform urban mobility? *Transp. Rev.* **40** 734–53

Cervero R and Kockelman K 1997 Travel demand and the 3Ds: density, diversity and design *Trans. Res. D: Transport and Environment* **2** 199–219

Corais F 2016 Leituras Morfotipológicas e os IGT's, *in PNUM—Rede Lusófona de Morfologia Urbana* (Guimarães: PNUM) pp 327–43

Corais F 2021 *A Cidade a 'Caminhar' para 2050. Braga como Laboratório para um Sistema Urbano Resiliente e Sustentável (Projeto Tese)* (Universidade do Minho)

Daugėlaitė A and Gražulevičiūtė-Vileniškė I 2018 Urban acupuncture in historic environment: research of analogues *J. Sustain. Archit. Civil Eng.* **23** 5–15

de Solà-Morales M 2008 *De cosas urbanas* (Barcelona: Editorial Gustavo Gili) 8425222605

Gehl J 2011 *Life Between Buildings: Using Public Space* 6 (Washington, D.C.: Island Press) 9781597268271

Gehl J and Svarre B 2013 *How to Study Public Life* (Washington, D.C.: Island Press) 978-1-61091-525-0

Hoogduyn R 2014 *Urban acupuncture Revitalizing urban areas by small scale interventions* Blekinge Tekniska Högskola http://bth.diva-portal.org/smash/get/diva2:832526/FULLTEXT01.pdf

Lerner J 2003 *Acupuntura Urbana, Investigaciones geográficas* (Rio de Janeiro: Editora Record) 8501068519

MPT; Câmara Municipal de Braga 2017 Projeto 'Área mais'—zona Quinta da Fonte

Município de Braga; Quaternaire 2016 *Projeto de Alteraçãi da ARU de Braga Sul e de Delimitação das ARU Braga Norte e Braga Nascente.* https://ru.cm-braga.pt/0101/reabilitacao-urbana/aru-zona-consolidada/enquadramento

# Chapter 6

# Tactical urbanism: collective interventions in urban public spaces

**Matías Cardoso, Daniel Orellana and M Augusta Hermida**

Tactical urbanism has emerged as an umbrella concept that uses rapid, low-cost experimental urban interventions to quickly transform and improve public space. It can be seen as a prototype for flexible intervention that allows for adjustments and improvements before the implementation of a long-lasting project. This idea has gained traction worldwide during the last decade, fuelled by the need to redistribute the public space, mainly in urban areas where the car-centric model predated the streets and displaced other street activities that are part of urban life. In this chapter, we revisit the fundamentals of tactical urbanism focused on active mobility, review a wide range of interventions worldwide, and propose a roadmap for successful interventions focused in Latin American cities.

## 6.1 Introduction

During the rapid urban expansion that has taken place since the mid-20th century, most Latin American cities have witnessed an increasing population migration from rural areas. As a response, urban planners in the region embraced the zone-based paradigm, assigning different parts of urban life—work, home, and leisure—into separate geographic zones, increasing transportation demand, and reassigning large proportions of public space for the region's growing number of private cars. The irregular topography of the Andes, together with the market force of land value speculation, deepened the sprawl, extending the urban footprint into hills and valleys, intensifying the demand for motorised mobility and increasing commuting times. This car-centred urban model has proved to be detrimental to the environment and urban populations, due to: high rates of traffic-related deaths and injuries, the increased risk of heart disease and diabetes due to the lack of physical activity, increased rates of asthma and respiratory disease due to air pollution, greenhouse emissions, the diminished quality of public spaces, and a weakened sense of community and attachment at the neighbourhood level.

doi:10.1088/978-0-7503-3971-1ch6

In recent years, it has become clear that cities require a change of paradigm in mobility. This new paradigm includes shortening distances by enhancing land-use diversity, promoting active mobility modes, such as walking and cycling, and improving the quality of public spaces, especially streets. This need has become especially evident during the COVID-19 pandemic, when public spaces have been among the safest places and have provided excellent opportunities to citizens for physical activity and psychological wellbeing. Consequently, the car-oriented street infrastructure prevalent in many Latin American cities is becoming outdated and requires significant transformation and adaptation to make it appropriate to fulfil the needs of 21st century urban life. In this new paradigm, the public space is reclaimed for the people, and cars step back to their utilitarian purpose instead of their current central role. This new paradigm involves different approaches and terms, such as 'complete streets', 'streets for all', and 'pacified streets', among others.

Such interventions, however, may require an important financial investment that many intermediate and small cities are unable to make. Faced with the need for short-term solutions to transform public space, many municipalities, citizen associations, and collectives have devised low-cost approaches to the design and implementation of medium- and short-term interventions to improve the quality of urban spaces with relatively little effort. These temporary and focused projects may be conceived and implemented by one or many actors, such as local governments, citizen groups, neighbourhood associations, or even individuals. The general process of these interventions is usually known as *tactical urbanism.*

During recent decades, fast urbanisation has overwhelmed urban planning departments, causing public spaces to grow in an informal and unorganised way. In this context, tactical urbanism has been widely used by citizens willing to transform their neighbourhoods into more livable and vibrant built environments. This approach enables citizens to collaborate actively in the construction and improvement of their neighbourhoods. Tactical urbanism is a powerful tool for communities with informal economies, and it usually acts as a connector between informality and a much-needed formality in urban planning processes.

This chapter proposes a roadmap for tactical urbanism interventions (TUIs) focused on mobility solutions in Latin American intermediate cities. First, we revisit some fundamentals of tactical urbanism; we then review more than 50 experiences worldwide to extract and summarise the main features and commonalities; finally, we present our five-stage proposal for the planning, design, and implementation of successful TUIs.

## 6.2 Fundamentals of tactical urbanism

The Merriam-Webster Dictionary defines *tactics* as 'the art or skill of employing available means to accomplish an end'. In this case, the term implies a short-term action capable of changing a situation, in this case, some urban feature. Formally, tactical urbanism emerged in 2010 from the Next Generation of New Urbanists movement (Lydon 2016). However, almost every city around the world has examples of citizen-driven practices of urban transformation. During the last decade, tactical urbanism has been used as an umbrella term to refer to low-cost, temporary urban interventions that improve local neighbourhoods (Pfeifer 2013). Some examples of TUIs include transforming an empty lot into a playground, implementing temporary cycle lanes, or improving a pedestrian crossing with vivid colours, painting, and vegetation.

TUIs usually employ low-cost materials to test design solutions in the real world with the potential to improve the quality of public life of a neighbourhood. They can be seen as a short-term prototypes that yield relevant information for long-term planning and the design of urban features. They also represent an instance of public participation by individuals and collectives, allowing them to express their needs and to demonstrate their views about public spaces. The collective action of tactical urbanism allows neighbours and users to reengage with their neighbourhood's public life, strengthen their relationships, and reinforce the sense of community. A TUI represents a materialisation of a collective vision imprinted in the public space. This citizen participation process allows residents to interact with the proposal and experience its benefits or problems before it becomes a permanent project.

Moreover, the highly localised solutions resulting from TUIs for public spaces can enrich the urban planning process at the city scale, since they provide lessons and collective learning that planners and decision-makers can incorporate into their approaches. The synergy between long-term planning and light, low-cost, rapid interventions constitutes an effective formula, not only to validate long-term ideas but as a powerful tool with which to articulate and activate citizen involvement in relevant issues that have a positive impact on their quality of life. Moreover, an increasing number of local governments are gaining interest in tactical urbanism as a mechanism with which to reignite public participation and reengage public discussion aimed at more significant urban transformations.

In the context of Latin America, many cities face similar challenges regarding social equity and inclusion, urban segregation, and differential investment in public space, usually focused on high-value neighbourhoods or touristic areas. Nevertheless, public participation in building and maintaining public spaces and collective goods and assets was fairly common in the region until a few years ago, and it still persists in some areas. Such is the case of the *minga* or *minka*, a grass-root organisational form in the Andean communities based on collective voluntary work and focused on building, fixing, or maintaining some collective asset, such as a school, park, or community water system. The *mingas* that aimed to create or maintain public spaces were commonplace until recently in Andean cities as a community response to the lack of investment from local governments. This practice, however, declined during the last decades, and with it, the shared responsibility and the community sense of attachment to public spaces faded away. Tactical urbanism can be seen in Latin America as a reimagined form of urban *minga* that recovers this historic collective practice.

TUI focused on improving conditions and infrastructure for cycling and walking has become a widespread practice in many Latin American cities, both as a pragmatical approach to explore and test design alternatives and as a ploy on the part of civil organisations and collectives to reclaim urban space for residents and diminish the local impact of heavy motorised traffic. When correctly applied, this approach can quickly transform the built environment to provide places to relax, walk, and cycle safely for people of all ages. In addition, it usually promotes local economic development and raises awareness about the consequences of motorised vehicles on the urban environment.

Due to their nature, mobility-based TUIs may arouse conflicting reactions in different groups. Often, the most vigorous opposition comes from those who use these spaces as 'pass-through' streets in their daily commute and have no attachment to the place, whereas residents usually support and promote the process. Nevertheless, this kind of reaction is desirable, since it is a valuable indicator of the diversity of perspectives regarding public spaces. Also, they constitute a test of the water for more extensive, permanent projects that usually represent significant investments. Short-term TUIs aim to experiment and demonstrate different alternatives to public space design and do not necessarily represent a one-size-fits-all definitive solution. After being tested and accepted by the community, they serve as the basis for long-term urban projects. Even though not every TUI results in a permanent project, they encourage the community to discuss public space values, public life, and the different ways to improve their neighbourhood.

Despite their vast potential, and probably due to their relaxed nature, most TUIs are not fully documented and analysed, and their actual impact is poorly understood. For example, an intervention aimed at improving pedestrian safety at an intersection would require an estimation of the change in traffic collisions or assessment of people's perception of safety. However, if no prior and post-intervention data are collected and compared, it is difficult to evaluate the effectiveness of the intervention.

## 6.3 Tactical urbanism around the world

We reviewed 52 mobility-related TUIs from 14 countries reported in the scientific literature, white papers, technical reports, and institutional documents. Nearly 45% of the cases were located in North America, another 45% in Latin America, and 10% in Europe and Asia (figure 6.1). These figures represent a publication bias more than an actual geographic pattern, but they illustrate a renewed interest in revitalising public space in the Americas.

For each intervention, we documented a set of characteristics including: the *name* of the intervention; *images* illustrating the main features; a *short description* of what the intervention was about; the *aim* or goal, according to the reporting document; *keywords* used to find the intervention in search engines; the location of the TUI (*country* and *city*); the *duration* and *dates* of the intervention; the *scale* of the project and its area of influence; the *organisations* involved in the project and their type (public, private, civic); and a *link*, i.e. a URL leading to the source of the information. Some of these characteristics are summarised in table 6.1. A full table with all the details and images can be found at http://llactalab.ucuenca.edu.ec/piesypedales.

Although the exploratory nature of this review does not allow for a comparative analysis, it provides some interesting insights (figure 6.2). Nearly 33% of the interventions aimed to create *pop-up* cycleways and cycle lanes, a kind of intervention

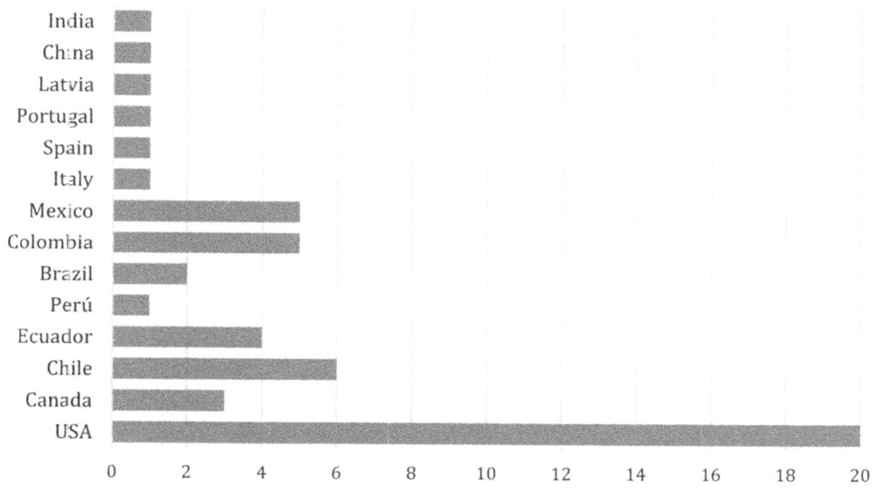


**Figure 6.1.** Numbers and countries of TUIs included in the review. Source: the authors.

that usually requires simple colour painting and, in more advanced interventions, physical objects to separate the bike lane from motorised traffic. About 41% of the TUIs explicitly focused on reclaiming public space for pedestrians. One of the most successful 'tactics' consists of creating temporary parklets in parking spaces and revealing the amount and diversity of activities that a space occupied by a single car can accommodate. Other common interventions include pedestrianisation, where motorised access to a street segment is temporary or the street is permanently restricted to pedestrians and non-motorised vehicles, and sidewalk extensions to diminish the crossing distance for pedestrians at an intersection. Also, 10% of the interventions followed the 'complete streets' approach, aiming to adequately redistribute the space among all modes and activities in the street.

## 6.4 Types of mobility-focused TUI

In this section, we describe the main types of TUI that are focused on mobility. For each type, we provide a brief explanation and design considerations and also summarise the recommendations provided in guides and handbooks (Lydon 2016; Trailnet 2016).

### 6.4.1 Sidewalk extension

This consists of expanding the space of the sidewalk, occupying parking spaces near intersections to narrow the lanes, reduce the speed of vehicles, and shorten the crossing distance for pedestrians. This modification gives pedestrians extra space for crossing intersections, while forcing vehicles to slow down when approaching and making turns. By being separated from parking lots, these extensions give pedestrians greater visibility to decide whether to cross or not and make them more visible to oncoming vehicles.

**Table 6.1.** Summary of TUIs reviewed during this study. Source: the authors.

| Name | Country | City | Focus | Type | Duration | Year | Scale | Institutions |
|---|---|---|---|---|---|---|---|---|
| Parklets in San Francisco | USA | San Francisco | Urban furniture | PPP | P | | Parking place | LG |
| Sunset Triangle Plaza | USA | Los Angeles | Pedestrian | PE | P | 2012 | Neighbourhood | PS + LG |
| Zhengtong Avenue redesign | China | Shanghai | Street reshaping | B | P | | Neighbourhood | LG |
| Paseo Bandera | Chile | Santiago | Pedestrian | PE | P | 2018 | Neighbourhood | LG + PS |
| Ocupa tu Calle | Peru | Lima | Pedestrian public space | PPP | P | 2015 | City | CO |
| Propuestas urbanas para Chile | Chile | Puerto Varas | Street reshaping | PE | P | 2016 | City | PS |
| Calle Rosa | Portugal | Lisbon | Pedestrian | PC | P | 2011 | Neighbourhood | PS + LG |
| Participative planning in High River | Canada | High River | Cyclists—Pedestrians | S | P | 2016 | Neighbourhood | CO |
| Stone Way North | USA | Seattle | Street reshaping | B | P | 2007 | Neighbourhood | LG + CO |
| Prospect Park West | USA | Brooklyn, NYC | Street reshaping | B | P | 2010 | Neighbourhood | CO + LG |
| The Porch (La Galería) | USA | Philadelphia | Pedestrian | PPP | P | 2011 | Neighbourhood | LG + PS |
| Crossings for cyclists | USA | Portland | Cyclists | PC | P | 2016 | City | LG |
| Urbanismo táctico en Barrio Conquistadores | Colombia | Medellin | Urban furniture | C | P | 2017 | Intersection | LG |
| Supermanzanas en Barcelona | Spain | Barcelona | Street reshaping | PE | P | 2008 | Neighbourhood | LG |
| Tactical urbanism in Times Square | USA | New York | Pedestrian | PE | P | 2009 | Intersection | LG |
| Urbanismo táctico en Barrio Conquistadores | Colombia | Medellin | Pedestrian | S | P | 2017 | Intersection | LG |
| Urbanismo táctico en Av. la Playa | Colombia | Medellin | Pedestrian | S | P | 2017 | Neighbourhood | LG |
| Estación Peatonal | Ecuador | Cuenca | Urban furniture | PPP | R | 2017 | City | LG |
| CicloRecreoVía | Chile | Santiago de Chile | Weekly road closure | B | R | 2005 | City | LG |
| Arica's bikeways | Chile | Arica | Cyclists | B | T | 2015 | Neighbourhood | CO |
| The complete street | Canada | Haliburton | Street reshaping | B | T | 2015 | Neighbourhood | CO |
| Mini-parque en Valdivia | Chile | Valdivia | Urban furniture | PPP | T | 2014 | Parking place | PS |
| Nagpada crossing | India | Bombay | Urban furniture | C | T | 2017 | Intersection | LG |

| | | | | | | | | |
|---|---|---|---|---|---|---|---|---|
| #5.80x2.15 | Mexico | Mexico City | Urban furniture | PPP | T | 2015 | One parking place | PS |
| Micro parque. Plaza del Mercado Restrepo | Colombia | Bogotá | Urban furniture | PPP | T | 2017 | One parking place | CO |
| Biscayne Green | USA | Miami | Urban furniture | PPP | T | 2017 | Neighbourhood | LG |
| Parklets en Ibarra | Ecuador | Ibarra | Urban furniture | PPP | T | 2016 | Neighbourhood | AI |
| Plaza Activa | Chile | Valdivia | Pedestrian | PE | T | 2014 | Neighbourhood | PS |
| Areas de transito calmo en Brasil | Brazil | Fortaleza | Urban furniture | PE | T | 2017 | One block | LG |
| Peatonalización en Sao Paulo | Brazil | Sao Paulo | Pedestrian | S | T | 2017 | Neighbourhood | LG |
| Urbanismo a escala humana en Bogotá | Colombia | Bogotá | Pedestrian | PC | T | 2016 | Neighbourhood | CO |
| Cebratón | Mexico | Puelba | Pedestrian | PC | T | 2015 | Neighbourhood | PS |
| Manos a la Obrera | Mexico | Mexico City | Urban furniture | C | T | 2016 | Intersection | CO |
| Cognición Urbana | Mexico | Queretaro | Pedestrian | C | T | 2016 | Intersection | CO |
| Cyclists of Roma | Italy | Roma | Cyclists | B | T | 2015 | Neighbourhood | CO |
| Urban intervention in Riga | Latvia | Riga | Cyclists—Pedestrians | B | T | 2017 | Parking places | PS |
| Frida | Ecuador | Quito | Urban furniture | PPP | T | 2016 | Parking place | PS |
| BikeNWA | USA | Bentonville | Cyclists | B | T | 2016 | City | PS + CO |
| San Marcos Code | USA | San Marcos | Cyclists | B | T | 2014 | Neighbourhood | PS + LG |
| Greenbelt pilot project | USA | San Francisco | Cyclists | B | T | 2014 | Neighbourhood | PS + CO |
| Market Street Connection | USA | Poughkeepsie | Urban furniture | B | T | 2016 | Neighbourhood | PS + LG |
| Complete streets demonstration | USA | Morgan Hill, CA | Street reshaping | B | T | 2014 | Parking place | PS + CO |
| Communities for life | USA | Atlanta | Cyclists—Pedestrians | B | T | | Neighbourhood | PS + LG |
| Plan BTV Walk Bike | USA | Burlington | Cyclists | B | T | 2017 | City | PS |
| North Minneapolis Greenway | USA | Minneapolis | Cyclists | B | T | | Neighbourhood | PS |
| Lincoln hub | USA | Chicago | Street reshaping | C | T | 2015 | Intersection | LG |
| 40th Street Parklet | USA | Oakland | Urban furniture | PPP | T | 2012 | Two parking places | PS |
| JC Walks Pedestrian Enhancement Plan | USA | Jersey | Street reshaping | S | T | 2017 | Intersection | PS |

(*Continued*)

**Table 6.1.** (*Continued*)

| Name | Country | City | Focus | Type | Duration | Year | Scale | Institutions |
|---|---|---|---|---|---|---|---|---|
| Demonstration in Islington Street, Portsmouth | USA | Portsmouth | Urban furniture | S | T | 2016 | Intersection | PS + LG |
| Tactical urbanism Workshop | Canada | Hamilton | Pedestrian | S | T | 2013 | City | PS |
| Mini ciclovia en Mexico | Mexico | Mexico City | Cyclists | B | T | 2018 | Neighbourhood | CO |
| Urbanismo táctico en Quito | Ecuador | Quito | Street reshaping | S | T | 2017 | Intersection | CO |

Type: PPP = Pedestrian plazas and parklets, PE = Pedestrianisations, B = Bikeways, PC = Pedestrian crossings, S = Sidewalk extension. Duaration: P = Permanent, R = Recurrent, T = Temporary. Institutions: LG = Local government, PS = Private studio, CO = Citizen organization, AI = Academic institution.

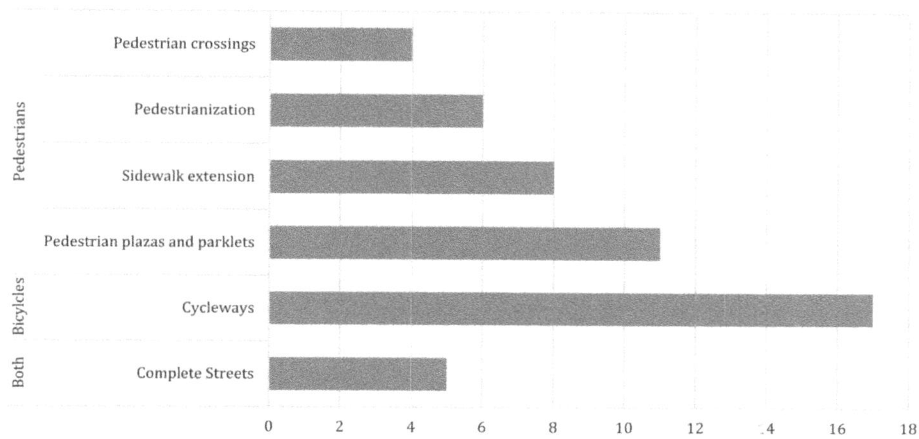


**Figure 6.2.** Numbers and types of TUI included in the review. Source: the authors.

*Design considerations and recommendations*: this intervention is most necessary where there is a large volume of pedestrians. It require parking to be available on the road near the intersection. Its implementation is easier if a pedestrian crossing already exists at the intersection. The width of a sidewalk extension must be at least the same width as an existing pedestrian crossing on the road. The extension of the sidewalk must be at least 30 to 60 cm narrower than the width of the parking spaces on the road. It is advisable to include barrier elements that limit the extension of the sidewalk and protect pedestrians and cyclists.

### 6.4.2 Pedestrian crossings

Safe and frequent crossings are necessary for more walkable streets. Well-designed pedestrian crossings spread out in places of greater demand can make walking a more attractive and safer way of getting around. To highlight pedestrian crossings with tactical urbanism, they can be painted with creative designs to stand out, so that drivers pay more attention when they approach them. Additional drawings can be added to the crossings or other road elements to reflect the neighbourhood's identity.

*Design considerations and recommendations*: the site considered for painting a pedestrian crossing must offer good visibility conditions to enable safe crossings. It is advisable to carry out these interventions on existing pedestrian crossings approved by the city traffic authorities. When working on wide or multi-lane roads, it is preferable to create or improve the existing islands in the middle of the crossing so that pedestrians can cross by sections. A pedestrian crossing must be at least 1.80 meters wide.

### 6.4.3 Pedestrian plazas and parklets

Park(ing) Day interventions arose in 2005 as an initiative of the Rebar collective in San Francisco (USA) to draw attention to the lack of green areas that the city suffered from. They consist of converting one or two parking spaces into a pocket park for a few hours or days. They are usually initiatives led by neighbours or civil activists and require very little investment. In 2008, a network called Park(ing) Day was launched to encourage this type of intervention worldwide. That same year, its popularity grew to the point of being one of the interventions carried out within the framework of the Biennial of Venice. Park(ing) Day interventions have become quite popular, with approximately 800 implementations worldwide each year (Coombs 2012).

*Design considerations and recommendations*: when the street still needs temporary parking spaces for vehicles, do not modify all available spaces. This type of intervention works better when combined with the work of local artists or activities that attract the public's attention. The squares and pocket parks in streets with little traffic and active facades open to the sidewalk are usually better perceived by the public. Use lines at least ten centimetres wide to delimit the pedestrianised space; in addition, use barrier elements such as posts and flowerpots to create a clear boundary.

### 6.4.4 Bikeways

Temporary bike lanes created only with paint or protected by flowerpots or dividers create new safe spaces for active mobility. Temporary bike lanes can be as short as three blocks, as they contribute to solving conflicting areas. They have multiple benefits, such as illustrating possibilities for long-term change, encouraging users to move by bicycle, and offering safety to those who already use their bikes.

*Design considerations and recommendations*: the road should be wide enough to include a bicycle lane without excessively hindering vehicular circulation. The lines that separate the bike lane from the road should be between 15 to 20 centimetres wide. Physical barriers should be separated by between two and six meters to prevent vehicles from parking on the bike lane, while allowing access to garages and inner parking.

### 6.4.5 Complete streets

Redesigning intersections to make them safer and more equitable for all types of mobility can be a powerful solution at heavy traffic spots. In many cases, this type of intervention can be combined with temporary public spaces or parklets. The redesign of an intersection requires knowing the volume and planning accordingly. In this type of intervention, the concept of complete streets can be applied to make roads safer and more equitable for all.

*Design considerations and recommendations*: intersections with enough space allow the implementation of changes that benefit all modes of mobility, since they help to create order and improve the flow. It is advisable to reduce the turning radius in minor intersections to limit the vehicle's speed and reduce pedestrians' crossing distance. The edge of a roundabout must be at least 4.50 meters from the closest corner of the sidewalk.

### 6.4.6 Pedestrianisation

It is possible to create different circulation, recreation, and resting places in zones underused by vehicles. These temporary public spaces can improve the economy and pedestrian mobility of a neighbourhood. The documented experiences indicate that they usually work better if they have a business and other activities nearby that can sponsor the intervention by offering a service or product in the new space. These spaces can make public life more attractive, facilitate pedestrian mobility, and test alternatives for long-term changes.

*Design considerations and recommendations*: its benefits are most noticeable in low-traffic or underutilised streets. They work better when there are active businesses nearby. It is advisable to intervene in this way in areas with a significant potential for pedestrian flow. Pedestrianised space must be clearly marked with barrier elements and different pavement colours. It is better to keep the space compact, so that the activity that takes place in it does not spread out too much. Take care that the elements used to separate the plaza from the road do not reduce the visibility of pedestrians or prevent them from crossing the space freely.

## 6.5 Methodological approaches

Among the cases documented in the review, only a few included information about the methodology or the process used to plan, design, and implement the TUI. Nevertheless, some organisations and institutions have published handbooks and manuals for tactical urbanism. This section summarises two of the most common approaches: *Quick Build for Better Streets* from People for Bikes and *Action-Oriented Planning* from Making Cities for People (Gehl).

### 6.5.1 Quick build for better streets

The Quick Build for Better Streets (QBBS) methodology was published in 2016 by the American association *People for Bikes*. Although it is specifically designed to build new bike lanes, its authors argue that it can be used for projects of any kind in public spaces. Their approach is based on four assumptions: (a) the projects are led by the public sector, either the municipality or a government department, (b) they are installed approximately one year after the design process starts, (c) they are planned by considering that changes and adjustments can be made during their operating period, and (d) they are built with materials that allow changes and adaptations during operation.

The approach encompasses four stages: (a) *Demonstration*, focused on implementing a real-world proof-of-concept with very few resources and simple materials, intended to test an idea and illustrate the overall project goal. This stage has maximum flexibility and allows for any necessary change. (b) *Pilot*, which aims to investigate how the intervention develops under different conditions and to collect data and information. It also allows an investigation of unforeseen positive or negative impacts and for the project to be adjusted accordingly. (c) *Provisional*, intended to fine-tune the final design. Finally, (d) *Definitive*, in which the final project is implemented with long-lasting materials and elements. Each stage takes incrementally longer and is 'harder' in the materials and elements used. This logic allows for more flexibility during the earlier parts and more durability later (table 6.2).

It is worth emphasising that although the QBBS approach focuses on government-led projects, it considers and encourages public involvement.

**Table 6.2.** Quick Build project process. Source: Authors, adapted from Orcutt and Andersen (2016).

| | Stage | | | |
|---|---|---|---|---|
| | Demonstration | Pilot | Provisional | Definitive |
| **Image** |  |  |  |  |
| **Duration** | 1 day–1 month | 1 month–1 year | 1 year–5 years | 5 years–50 years |
| **Organiser** | Any person or group | Authorities with the participation of users and collectives | Authorities with the participation of users and collectives | Authorities with the participation of users and collectives. |
| **Permits** | Sanctioned or unsanctioned | Always sanctioned | Always sanctioned | Always sanctioned |
| **Materials** | Low cost and low durability | Relatively low cost and medium durability | Low to moderate cost, designed to balance design flexibility and maintenance costs | Expensive materials, high durability, and little flexibility |
| **Public involvement** | Organised, built, and maintained by users, neighbours, or other actors of the city. | Contributions from the public, but built and administered by the authorities | Contributions from the public, but built and administered by the authorities | Contributions from the public, but built and administered by the authorities |
| **Flexibility** | High. The project can be adjusted on the run and is removed when it reaches the end of its useful life. | High. The project can be adjusted on the run and is removed when it reaches the end of its useful life. | Moderate. Minor adjustments can be made, but the project is expected to remain on site. | Low. The project is considered a significant investment in the city, and no changes are expected once installed. |
| **Data collection** | Recommended | Always | Always | Always—these data are used to plan future investments and projects. |

### 6.5.2 Action-oriented planning

Action-Oriented Planning is an approach proposed by Gehl Studio (Risom *et al* 2016) using tactical urbanism as a tool to test a proposed design and analyse its impact. The authors argue that citizens use a pilot project in often unimaginable ways, unintended by the designers. Thus, monitoring, evaluation, and reinvention are crucial to determine the project's success. In this approach, tactical urbanism serves as a tool for assessing public spaces' options and understanding how they impact human behaviour. The information obtained from the evaluation provides inputs for decision-making processes as part of the overall planning or individual projects' designs. Although focused on small interventions, the broader objective of the proposed methodology is directed at city leaders, who must institutionalise people-centred approaches to link government to civil society. By using people-centric metrics and tactics, it is possible to cultivate a better quality of life for all residents and allows citizens to become visible in the data collection of every institution in the city. In this way, the built environment becomes more habitable and adaptable to the human scale (Risom *et al* 2016). At the core of Action-Oriented Planning is a three-stage cycle (*measure*, *test*, *refine*) that is iterated to produce a final project (figure 6.3). The stages are briefly described below.

*Measure*: Collecting information and data to build a baseline for a project is crucial. People-centric metrics allow the collection of evidence to support change, creating a buzz for projects, and persuading sceptics to participate. The data can also reveal patterns previously unnoticed by designers. The information can be collected by combining observations, surveys, various forms of participation, and collecting quantitative data on the physical composition of public spaces. The performance of a prototype can be measured using several criteria, which serve as the basis for generating and comparing adjustments that lead to a better design. Although each project requires the definition of its evaluation parameters early in the process, as the project progresses, these parameters may be modified.

*Test*: The TUI can be seen as a test that evaluates a set of design criteria and understands the perceptions and interactions of the different stakeholders. Motivating more people to participate early in the planning process creates more inclusive and practical engagement. Also, inviting citizens to test initiatives directly before making a significant investment can reduce the risk of failure and helps to ensure that investments are used as effectively as possible to maximise positive impact.

*Refine*: The prototype nature of TUIs allows the exploration of many possibilities and quick experimentation with short-term transformations that can improve the initial idea and evaluate its performance. The modifications can be small or big, but are always a product of the learning process.

This cycle constitutes a feedback loop that produces a final project. Using the lessons of the early pilots to improve the design of a permanent intervention offers a greater chance of success.

## 6.6 A roadmap for tactical urbanism interventions

In this section, we propose a methodological approach to mobility-focused tactical urbanism that is adapted to Latin American cities. This adaptation considers the

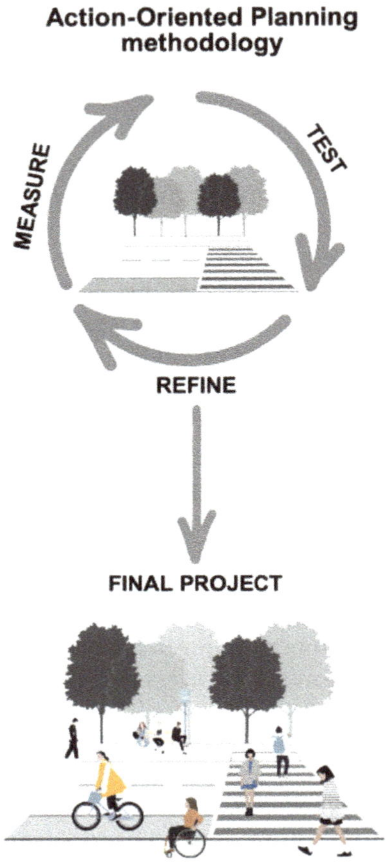


**Figure 6.3.** Action-Oriented Planning methodology. Source: the authors, adapted from Risom *et al* (2016).

prevalent lack of information at the street-level scale and several common problems related to car-oriented public spaces that characterise the urban areas in the region; nevertheless, it can be helpful to other regions and cities. The aim is to present a roadmap for practitioners, urban planners, and civic organisations looking for a guide with which to plan, design, and implement a successful TUI focused on improving people-centric public spaces. The proposal is firmly based on the lessons derived from the reviewed cases in the literature and our own experience of tactical urbanism.

The methodology combines several tools and methods to create a five-stage process for planning and organising evidence-based TUIs. Besides this chapter, we have created a resource handbook on tactical urbanism, which is available at http://llactalab.ucuenca.edu.ec/recursos.

### 6.6.1 Diagnostic

The diagnostic aims to understand the current situation of the space where the intervention will take place. This stage is crucial to generate an evidence-based intervention. The diagnostic must include three dimensions: (a) user behaviour, (b) user perceptions, and (c) streetscape features. We present tools and methods for each dimension, which may be helpful, depending on the aim and focus of the project.

#### *6.6.1.1 User behaviour*

Studying people's behaviour in a public space is an effective way to understand the interactions between people and a place. Based on direct observation, this reveals what people do, how long they stay, and where and how they move. The techniques for studying user behaviour are based on Gehl and Svaarre's 'How To Study Public Life' (Gehl and Svarre 2013), where interested readers can find more details. These techniques are flexible enough to apply to most cases and can be adapted to fulfil specific needs.

The first technique, *static activity mapping*, involves surveying what is happening in the place using a map. The objective is to find where people stay and what they do; if there are more men or women; if there are specific places where they are standing, sitting, or in any other position. The mapping should be repeated at different times and on different days to detect emerging spatial and temporal patterns. The information can be overlaid on aerial imagery or geographic information layers to explore the relationship between behaviour and space. The technique is simple in nature, requiring only a printed schematic plan that includes furniture, roads, vegetation, and other features. It is advisable to prepare a legend of easy drawing symbols representing the different activities. Also, different colours can be used to represent gender or age groups.

Another technique consists of *tracing pedestrian and cyclist movements* by drawing lines on a map to understand the direction, flow, and general spatial behaviour of pedestrians or cyclists in the area and to detect emerging patterns. It also reveals the most used paths, where people prefer to walk and if there is any difference between the routes chosen by men and women. The tracing time will depend on the number of people and can range from minutes to half an hour. In addition to the lines, information about the person's direction, age, or gender can be included by adding symbols or colours to the end of each line. Although the technique is not accurate enough to estimate volumes, it provides valuable information about the microbehaviour that supports design decisions in the later stages. This technique can be challenging in crowded places, in which case, it is advisable to select a random individual each time for tracing.

#### *6.6.1.2 User perceptions*

Whereas observing people in public spaces allows the study of their spatial behaviour, the analysis of perceptions seeks to understand people's motivations, attitudes and feelings about quality, safety, and comfort as they interact with the place. These results help us to understand how infrastructure problems affect people's perception, and where more work needs to be done to make a place more pleasant and safe.

One of the most common techniques is the *survey*. This is a form with sets of questions on different topics regarding the place and its characteristics. The focus must be on the main aspects related to the project, and the questions must be clear and direct and present closed options when possible. The questionnaires must be short and easy to answer to ensure that people can answer in a couple of minutes.

A good questionnaire must be validated for clarity and consistency using a pre-test. There are several guides and methodologies available for evaluating projects in public spaces. Nevertheless, it will be necessary to adapt them to, and validate them in, the new environment if they will be applied in a different geographic or cultural context. The use of pre- and post-intervention surveys allows longitudinal comparisons to be made, the results of which are beneficial in evaluating the impact of an intervention.

A third technique is the *pedestrian accessibility audit.* This tool was initially developed by Trailnet (USA) and adapted to Latin American contexts by the LlactaLAB Research Group. The adaptation also extended the original tool to include cycling. The instrument is conceived as a 'moving survey', i.e. a walking tour in the project area with the participation of residents, passers-by, planners and other stakeholders to collect their opinions and perceptions and assess how the built environment features affect walkability in terms of safety, aesthetics, and comfort. To apply the instrument, each participant receives a printed form for each section or block of the route on which to write down their evaluations and opinions. The researcher can video record the tour using a GPS-enabled camera to document and analyse the reactions and expressions, but they must consider that this may affect the participants' behaviour. Pedestrian accessibility audits are designed to stimulate discussion with citizens and raise awareness about how the built environment can encourage people to walk. This tool helps communities and intervention designers to understand what they need to change to promote active mobility.

Depending on the project requirements, additional quantitative and qualitative techniques can be used to study user's perception, such as interviews, focal groups, choice experiments, among others.

#### *6.6.1.3 Streetscape features*

The third component of the diagnostics is the walkability assessment of the project area through an audit of the street-scale features of the built environment that promote or hinder active mobility. There are several tools and instruments for assessing the walkability of the built environment: PEDS (Clifton *et al* 2007) aims to assess the quality of the built environment using surveys encompassing aspects of the built environment, pedestrian and cycling infrastructure, and road attributes. MAPS (Cain *et al* 2017) is a tool for auditing walkability attributes at the street scale which was originally designed for cities in the USA. eMAPS is a version of the original MAPS tool adapted for intermediate cities in Latin America and validated for three cities in Ecuador (Quezada and Orellana 2019). These instruments are designed to collect objective information about the built environment and produce synthetic indices of walkability through a scoring process.

The *eMAPS.ec* instrument is a comprehensive audit of the walkability of the built environment, including features of the street segments, sidewalks, intersections, facades, vegetation, among others. eMAPS is implemented as a GIS-based open-source toolkit that facilitates all steps of the walkability assessment. It includes a tool for the automatic generation of maps and forms for planning the logistics of the assessment, a mobile-based form for data collection and aggregation, a GIS plugin for data processing and representation, and an audit protocol. The results are automatically generated and include maps and tables with walkability scores for each segment and subscores for different dimensions.

Depending on the nature and goal of the intervention, other aspects can be studied. For example, volumes of pedestrians and cyclists are robust evidence of the impact of an intervention and can be obtained by systematic counts at specific segments or intersections. Air pollution and noise are usually crucial for residents and passers-by; therefore, pre- and post-intervention can help to visualise the positive impacts of a project that aims to promote active mobility. Moreover, low-cost sensors and devices are now capable of collecting and transmitting this information in real time, enabling educational and promotional activities.

### 6.6.2 Planning and design

Once the diagnostic is complete, the next step is to use the information obtained to clearly *define the project's objectives*, which must come directly from an understanding of the context and the needs of the people.

#### *6.6.2.1 Goals and indicators*

The chances of a successful project are increased when there is a concrete, realistic, meaningful, and verifiable goal, e.g. 'reduce the number of traffic injuries at this intersection by half', 'expand this sidewalk and generate physical separation from cars to improve the safety of children walking to school', or 'improve the safety and comfort of the bus stop by creating a pleasant parklet'. This primary goal may have (and usually has) have other positive impacts, such as diminishing pollution or improving the neighbourhood's identity. The planning must include clear and valuable metrics that can be used to establish benchmarks: punctual and feasible actions on site that allow the success of prototypes to be evaluated. The design must respond to the stated goal.

#### *6.6.2.2 Team and participants*

A TUI requires a team. The project must be approached from multiple perspectives, and requires experts with various skills in diverse areas. It is worth mentioning that the 'experts' are often the residents and individuals that use the place and are closely related to it. In this step, it is crucial to involve local leaders, decision-makers, public officials, residents, academics, advocacy groups, and any other person or organisation involved in the implementation. Lowering barriers to civic participation is critical to making people feel genuinely invited to be part of the planning process. Project promoters must actively look for creative ways to involve neighbours rather than just waiting for them.

#### *6.6.2.3 Design*

In the design step, the team meets to jot down ideas and produce sketches of the intervention. The team will increasingly improve the design through several iterations as they confront it with the data collected during the diagnostic phase and with diverse perspectives in participatory meetings. It is advisable to revisit the place of intervention several times to envision the implementation and to adjust the sketches to the real world.

Once the design is mature enough, the final design plans are produced. If technical expertise is available, rendered views and virtual reality scenes can be extremely helpful. The final plans must include measurements and the required materials and resources.

#### *6.6.2.4 Materials*

The materials used for tactical urbanism often rely on creativity and collaboration. Rather than just purchasing the required items, it is desirable to reuse existing elements and resources whenever possible. Items such as tape, paint, brushes, bollards, or concrete blocks can commonly be found in existing city supply inventories and can be reused for pilot projects.

#### *6.6.2.5 Communication plan*

There are multiple communication strategies that can be used to approach the inhabitants of a neighbourhood, to inform the about the possible impacts on the concurrence of people and local commerce that the intervention will bring. It is essential to create an outreach plan that communicates: what need/challenge does the project intend to address? How long will the intervention last? How will the project be evaluated? To what extent can project items be adjusted (or removed) in response to community feedback? (Lydon 2016).

#### *6.6.2.6 Organisation of logistics*

It is critical to have a flexible logistic structure for the implementation process. TUIs broadly differ from the typical urban project procedure, so expect to solve problems on the way and get creative during the project planning and implementation. Nevertheless, it is crucial to plan the time required to complete the implementation, the transport of materials, the security regarding traffic or robbery, alternative plans in case of bad weather, etc.

#### *6.6.2.7 Distribution of tasks and responsibilities*

One of the critical issues for a successful implementation is to have a precise distribution of roles and responsibilities among the participants. Due to the collaborative, multi-part nature of a TUI, and considering that many will be volunteers, the key responsibilities must be assigned accordingly. It is advisable to organise the activities into a matrix that includes the participants' names, tasks and responsibilities, times and schedules, people they must coordinate with, and who needs to be informed about any incident.

### 6.6.3 Implementation

#### *6.6.3.1 Environment and character*

The implementation will follow the established schedule during the planning. This step should be the most attractive and enjoyable part of the project. The promoters must engage with participants in playful ways and let the creativity flow. If the conditions allow, they should encourage the participation of children and other

**Figure 6.4.** Consecutive steps of the implementation. Source: the authors.

groups who are usually excluded. Music, puppets, or any kind of street performance and art helps to establish an enjoyable environment and attract public attention.

*6.6.3.2 Gradual implementation*

It is advisable to implement the pilot project on consecutive days, following an order that produces gradual changes that allow users to understand the changes and adapt smoothly: (1) horizontal signage, (2) vertical signage, (3) painted pedestrian areas, and (4) furniture and physical elements (figure 6.4). This progression will support the project by generating expectation, which needs to be considered in the communication plan.

*6.6.3.4 Adaptations and adjustments during implementation*

It is worth remembering that what is being built is a prototype and that it may be susceptible to changes and adjustments at any of its stages. Promoters must be ready to deal with potential problems that may arise and have a team that can make changes quickly and efficiently.

*6.6.3.5 Safety of volunteers and organisers*

During implementation, emphasis should be placed on the safety of volunteers and organisers. If the intervention includes areas previously used by motor vehicles, all participants must wear reflective vests during the implementation, monitoring, and dismantling of the intervention. Also, gloves and helmets might be necessary.

### 6.6.4 Evaluation

*6.6.4.1 Monitoring*

Continuous evaluation using qualitative and quantitative metrics serves to reveal the project's performance (Lydon 2016). The tools from the diagnostic can be used again to obtain comparable results and determine the impacts and their magnitude. The measurement of key metrics related to the main goal and secondary impacts after each modification of the prototype helps to develop criteria and guidelines for permanent designs (Risom *et al* 2016). Also, there are creative ways to collect perceptions and impressions. For example, a large board inviting people to write down their impressions and suggestions, or a hashtag to share on social media, will encourage discussion and debate.

It is paramount to understand that people's behaviour requires time to adapt. Usually, any change in the environment will initially spark a natural reaction until they become accustomed to the new configuration. One of the most common errors is to attempt to evaluate the success of an intervention after just a few days of implementation. Therefore, the monitoring must focus on the 'operational' microbehaviour, i.e., how people react to the modified environment. It is advisable to foresee strong reactions during the first days. Consider that most of the detractors will appear at this moment.

#### *6.6.4.2 Assessment*

After three or four weeks, the implementation will be mature enough, people's behaviour will have adapted, and new habits will be established. It is now time to evaluate the project. The assessment will focus on the stated goals of the design and will use the collected data. Both raw numbers and change rates help the impacts to be interpereted in different dimensions. Quantitative metrics are helpful, but personal information and perceptions allow for a deeper understanding and exploration of the nuances. Collecting reactions from local media and social networks will allow an understanding of diverse perspectives and motivations.

Generating charts and illustrations, presenting key figures, and producing before/after pictures are useful in order to demonstrate the project's impacts. They can be complemented by video or audio testimonies from the public and extracts from social media. At this point, the TUI becomes a learning experience that informs and improves the design process.

### 6.6.5 Final project

After evaluating the impact of the pilot project, the team will propose the final project for a long-term modification of the built environment. The project will detail which elements of the temporary intervention were successful and can, therefore, be replaced with better, long-lasting materials. It is advisable to maintain the same logic and character of the TUI, since the experience probably reinforced the sense of attachment and identity for residents. Also, there might be elements of the intervention that were unsuccessful and should be removed or redesigned. The local authorities will have had enough time to plan and secure financial resources to build a project that has already demonstrated its usefulness, and the TUI will become a long-term transformation of the urban fabric. Figure 6.5 illustrates an example of TUI by the authors in Cuenca, Ecuador.

The final task is to promote and publish the results and learned lessons (which will be extremely useful for other groups and cities) and to build up a corpus of shared knowledge on tactical urbanism.

## 6.7 Limitations and further research

Although the proposed approach is oriented to mobility-related tactical urbanism in Latin American cities, the proposed approach is generic enough to be a useful roadmap for other kinds of intervention and region. We highlighted the importance of monitoring through data collection and analysis, since we believe that any kind of urban transformation must be evidence-based. Nevertheless, we recognise that this

**Figure 6.5.** Examples of TUIs in Cuenca. (A) A temporary bicycle lane, (B) an aerial image of the intervention, (C) a safe crossing, (D) a parklet and bus stop. Source: the authors.

approach might represent an significant effort for small-scale organisations and informal interventions, and we therefore recommend adapting it accordingly. Further research will include a comparison of similar cases in different cities in order to improve the proposed methodology.

## 6.8 Tactical urbanism and the COVID-19 pandemic

The COVID-19 pandemic generated a set of unprecedented conditions for cities around the world. Mobility restrictions and disease transmission associated with mass public transport pushed thousands of people into rediscovering walking and cycling as mobility modes. Also, following the stay-at-home periods, people become more aware of the importance of ample and safe open public spaces where risk is at the lowest levels. In this context, the pandemic offered an unexpected opportunity to accelerate the implementation of new ideas for improving public spaces that reduce the risk of transmission, reclaim space from motorised vehicles, and diminish the inequality of inaccessibility to urban facilities. TUIs became the best way to implement rapid change with readily available materials that could be designed with a long-term vision.

These bottom-up approaches allowed fast adaptations, and during this time, citizens were more accepting of changes in their streets. During 2020, examples of TUIs popped up around the globe. In London, Berlin, Bogotá, Philadelphia, Minneapolis, Vancouver, Mexico City, and Milan, paint and traffic cones were used to create bike lanes.

In Dublin, parking places and loading bays were reclaimed in the city centre to provide more space for pedestrians. New Zealand created a nationwide tactical urbanism fund for emergency bike lanes and footpath widening at a national level. Besides reducing the risk of transmission, these mitigation strategies have also encouraged citizens to opt for more sustainable transportation modes, resulting in benefits in terms of improved air quality and health and reduced traffic noise and accidents.

These adaptations using tactical urbanism offered an excellent opportunity to collect data that will validly support permanent implementations once the pandemic is over. These projects can be considered as pilot programs that could involve the community in the selection of future permanent infrastructure. In this context, the methodologies and tools proposed in this chapter offer an easy and low-cost way to carry out the data collection and evaluation.

## Acknowledgement

The research presented in this chapter is part of the project 'Pies y Pedales 2', founded by the Research Directorate of the University of Cuenca. The authors express their gratitude to Caridad Pineda for her valuable contribution to the methodology of tactical urbanism presented in this chapter.

## Bibliography

Cain K L *et al* 2017 Developing and validating an abbreviated version of the microscale audit for pedestrian streetscapes (MAPS- abbreviated) *Journal of Transport and Health* **5** 84–96

Clifton K J, Livi Smith A D and Rodriguez D 2007 The development and testing of an audit for the pedestrian environment *Landscape and Urban Planning* **80** 95–110

Coombs G 2012 Park(ing) day *Contexts* **11** 64–5

Gehl J and Svarre B 2013 *How to Study Public Life* (Washington, DC: Island Press)

Lydon M 2016 *Tactical Urbanist's Guide to Materials and Design. Version 1* (The Street Plans Collaborative) pp 1–132. http://tacticalurbanismguide.com/guides/tactical-urbanists-guide-to-materials-and-design/

Orcutt and Andersen 2016 http://b.3cdn.net/bikes/675cdae66d727f8833_kzm6ikutu.pdf

Pfeifer L 2013 *The Planner's Guide to Tactical Urbanism* https://reginaurbanecology.files.wordpress.com/2013/10/tuguide1.pdf

Quezada A and Orellana D 2019 *e-Maps.ec: Herramienta de evaluación a microescala de ambientes peatonales* https://llactalab.ucuenca.edu.ec/recursos/

Risom J, Merker B and Muessig A 2016 *Planning by Doing. Gehlinstitute.org* https://gehlinstitute.org/wp-content/uploads/2017/02/20160301_Planning-by-Doing_print-1.pdf

Trailnet 2016 *Slow Your Street—A How to Guide* https://trailnet.org/our-work/planning/lending-library

**IOP** Publishing

Sustainable Urban Development

Topics, trends and solutions

**Luís Bragança, Cristina Engel de Alvarez and Luisa F Cabeza**

# Chapter 7

# Seeding picoscale solutions for social macro goals: complex thinking in projects for vulnerable communities

**Andrea Naguissa Yuba, Gilfranco Medeiros Alves, João Onofre Pereira Pinto, Juliana Trujillo and Karina Trevisan Latosinski**

This chapter presents lessons learned from two experiences of implementing socio-political sustainability strategies to achieve environmental and economic sustainability. The projects were carried out in rural communities in Guinea-Bissau and Jamaica, which are both vulnerable to climate change. Each experience, developed from a systemic perspective and under two frameworks (the water–food–energy nexus and 'Appropriate Technology'), is described in four phases (problem characterization, solution design, solution pre-evaluation, and implementation), with the significant findings made by the team in terms of challenges, rewards, and essential conditions for development. Both deal with environment-friendly solutions that solve the problems of water and energy provision. Based on complex thinking, the experiences are analyzed from the social inclusion, participation, empowerment, and learning theory viewpoints. Finally, the insights gained by the team are highlighted: communication requirements, the level of dependency of industrialized solutions, budget constraints, training, testing, and planning recommendations, the diversity of interaction with the communities, and partners' skills. As a result, future actions can exploit these experiences and contribute to the success of subsequent projects.

## 7.1 Introduction

The goals of social sustainability are commonly referred to in association with those of the economic dimension in an intricate relationship of cause and effect, something like 'wealth brings health'. This implied relationship holds true, at some level, for basic human needs (to end hunger, to offer medical care/medication/vaccines, to provide sanitation, drinking water, education, and employment), but that is not all.

doi:10.1088/978-0-7503-3971-1ch7

For life in society, there are more needs than just the economic ones. These needs also impact people's health and wellbeing, especially in the areas of accessibility, their sense of safety, equality, empowerment, and provision of opportunities. They could be linked to other dimensions, such as politics and culture.

Some of these needs cannot be provided by an external entity and they need to be achieved by the people themselves. In vulnerable and/or remote communities, carrying out these 'providing' actions (which are frequent, expensive, and do not generate improvement) is not always possible, which raises the challenge of generating 'enabling' actions. And it is up to the people who are 'enablers' to design ways to give the community a little support through this development process, while the enablers themselves learn from the process as well.

This chapter presents two of these cases, and although the peculiarities of each project are shown, both have the following in common: the issues that arise from the uncertainties of climate change, low environmental impact solutions, cost reduction, ease of appropriation, and low-maintenance solutions. These projects were implemented in Guinea-Bissau and Jamaica, countries in two different continents, because they originated from two distinct calls for proposals (CFP), for target communities established by the sponsors.

In each of these projects, the approach was similar, but depending on the existing conditions in each case, different results were obtained; it is therefore important to discuss these, so that future projects with the same intentions can use the strategies and lessons learned as a reference.

## 7.2 The frameworks merge

A social sustainability project's success may be measured by the effectiveness and longevity of the solution. The effectiveness may be determined by looking at the way in which a problem is solved, while longevity depends on the solution's maintainability. In the search for these features, different approaches may be used. For the two cases described in this chapter, the solutions were based on two frameworks: (i) the water, energy, and food security nexus—aiming at effectiveness; (ii) appropriate technology—aiming at longevity as well as low cost. Although these two frameworks are widely known, they are briefly described in this section for the sake of completeness.

### 7.2.1 Framework 1: the water, energy, and food security nexus

Social inclusion requires a quality of life that, at a minimum, includes food, water, energy, education, health, and housing. Among those, food, water, and energy stand out because they are required in an urgent and continuous fashion. Since the 1980s, this has been one of the factors that has led the United Nations to develop studies of the integration of the management and policies of these sectors, aiming for more sustainability, and therefore, effectiveness, in the search for the green economy. This integrated approach was called the 'water–energy–food nexus' (WEF-Nexus).

This subject became more popular after the Bonn2011 Conference, which had the theme of 'The Water, Energy and Food Security Nexus: Solutions for the Green

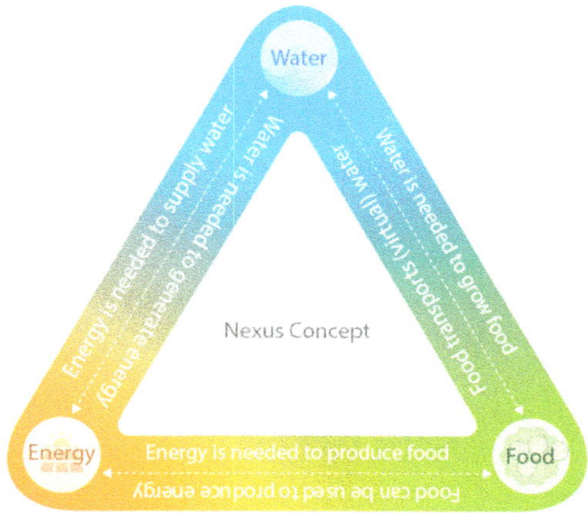


**Figure 7.1.** The water, energy, and food security nexus concept (adapted with permission from Hoff 2011).

Economy.' At that conference, the paper 'Understanding the Nexus—Background paper for the Bonn2011 Nexus Conference' by Holger Hoff (2011) was extensively discussed in order to obtain a full understanding of the concept and the possible impacts of this nexus. As presented in that paper, the concept of this nexus can be summarized in figure 7.1.

The water–energy interdependency summarized in figure 7.1 shows that, in one direction, water is necessary to produce energy. For instance, water is necessary to process and refine fossil fuels, for hydro-generation and biofuel production, among others. In the other direction, energy is needed to extract, treat, and transport water for human, agricultural, or industrial use. In the water–food interaction, water is one of the main inputs for food production. As a rough approximation, it is estimated that, on average, the production of one calorie of food requires one liter of water. The energy–food interaction has increased significantly as a result of agricultural mechanization, irrigation and fertilizers, and food processing, which requires energy as one of its main inputs. Energy production, in its turn, may use a large amount of land for fossil fuel mining, crop production for biofuels, etc.

The nexus approach was primarily proposed for macroscale management across the water, energy, and food sectors. In that sense, the purpose was for this nexus to be a framework that could be seen as a kind of decision support model for policymakers. So far, that is largely being used to aim for sustainability and move toward a green economy more rapidly.

As mentioned before, the WEF-Nexus approach was proposed for decisions/projects on the macroscale. However, nothing prevents the same framework from being applied at the picoscale. In this sense, the projects implemented in Guinea-Bissau and Jamaica used the WEF-Nexus framework. The main problem addressed was food production, in a horticultural garden and small coffee farms. But, to increase the food productivity and hence the production, it was necessary to have irrigation. Energy was necessary to extract water from wells and distribute it. For that, solar energy, harvested through solar panels, was used. So, the interdependency addressed in the WEF-Nexus framework was applied here with the aim of reaching social sustainability.

### 7.2.2 Framework 2: Appropriate Technology

In vulnerable communities, longevity, low cost, and the potential for replication are some of the most desirable features of a project. These features set a challenge for a project's proposers/implementers. A project's longevity depends not only on the design and the materials used, but also on its maintainability, i.e. how it is taken care of, and how often it is repaired when necessary. The cost depends on the design itself, including the materials, the workforce, and the equipment/parts used for the execution. Finally, the potential for replication relies on the design, the cost, and the transfer of know-how or technology to the community. Considering such challenges and requirements, one of the most promising frameworks for reaching success is the so-called Appropriate Technology (AT)[1].

One of the definitions of the word technology is: 'the skills, methods, and processes used to achieve goals', while one of the meanings of the word appropriate is: 'suitable or proper in the circumstances'. The combination of these words resulted in the framework named Appropriate Technology, and one of the best definitions that matches its origin was presented by Hazeltine and Bull (1999): 'any object, process, ideas, or practice that enhances human fulfillment through the satisfaction of human needs' and 'it is compatible with local, cultural, and economic conditions (i.e., the human, material and cultural resources of the economy), and uses locally available materials and energy resources, with tools and processes maintained and operationally controlled by the local population'. Figure 7.2 presents the concept of Appropriate Technology.

Inspired by such a framework, the design and implementation of the projects presented in this chapter had the following firm goals: the projects should use local knowledge/know-how and local low-embodied-energy materials (preferably renewable, non-hazardous products) and they should be low cost, easy to build and to maintain, and durable.

One important aspect of trying to apply the AT framework is that in some solutions, parts of the system require exotic technology. For instance, for irrigation projects using solar energy, the solar panels, inverters, and pump technologies are not local, even for most developing countries, therefore, for underdeveloped ones, they are certainly not. Thus, another challenge is how to implement an AT-based project if these types of equipment/parts are necessary. This then introduces the concept proposed by the authors, namely, the AT project index. This index measures how much of the project has AT content. Since the quantitative definition of such an index is under development, a brief and qualitative description is presented.

In designing a solution, some parts/types of equipment have no other choice but to use exotic technology. However, there are parts of the solution that can be based

[1] In this chapter, capital letters in 'Appropriate Technology' are used to denote the concept of AT, and lowercase letters denote a solution designed using AT.

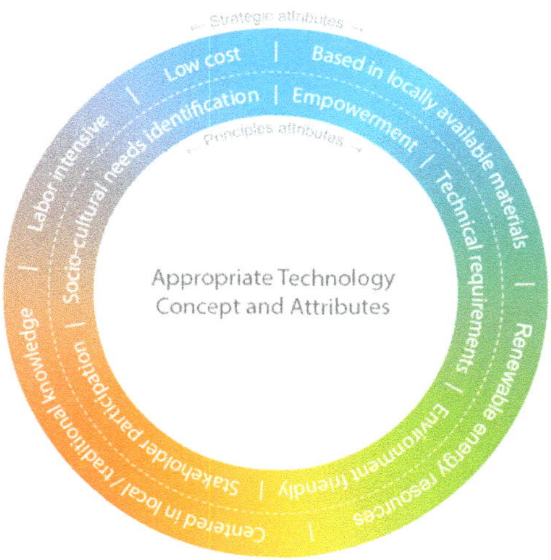


**Figure 7.2.** Appropriate Technology concept and attributes (adapted from Patnaik and Bhowmick 2019).

on local materials, knowledge, and workforce. Hence, the best design, in the AT sense, is the design that maximizes such local operability. For example, in a solar energy irrigation project, the whole construction project including the reservoir, the tower for the reservoir, and the hydraulics should be designed and implemented while aiming to increase the AT index. The two cases presented in this chapter made achievement of the highest AT index one of their goals.

## 7.3 Projects' untangling

### 7.3.1 Solar-energy-based irrigation project in Campada Maria, Guinea-Bissau

This project presents the development and installation of a water pumping pilot using photovoltaic solar energy for irrigation in a small collective horticultural farm in Guinea-Bissau, a country with a human development index (HDI) of 0.480–175th out of 189 countries classified by the United Nations Development Programme (2020).

In the Republic of Guinea-Bissau, on the West African coast, the primary sector is the mainstay of the economy. There, small horticultural farms are an important means of obtaining an income, which results in social inclusion and adds food security.

However, the hot climate typical of tropical regions, the dry seasons, and the lack of irrigation facilities diminish productivity and limit crop diversity. In many cases, the irrigation system does not exist, because an electric power grid with the capacity to install motorised pumps is not available; this is the reality in most of the country, due to the lack of infrastructure.

For many families, horticultural gardens and crops are the only source of income for women in the interior of the country, where, besides the economic issues, a male-centered society is prevalent.

**Figure 7.3.** Women's daily heavy-duty routine of collecting water by hand from wells.

The economic and technological limitations also contribute to a cultivation system that is still quite rudimentary, where irrigation is done with watering cans, and the collection of water from the wells is manual (figure 7.3). These activities are delegated to women, who, with the help of children, are responsible for feeding their families and for other domestic activities, which constitute a large workload. Thus, in a country with general poverty, the fragility of the State extends to agriculture, due to the scarcity of investment in agricultural equipment and materials and the lack of technological development and technical support for farmers.

The project was carried out by a partnership between UFMS—the Federal University of Mato Grosso do Sul, Brazil (conception of the solution), IBAP—the Institute for Biodiversity and Protected Areas, Guinea-Bissau (logistical and technical support and intermediation with the targeted community), CNPq—the National Council of Scientific and Technological Development, Brazil (financing), and a Brazilian technical support network of structural, hydraulic, materials, and geological researchers. The ideas that preceded the project implementation, supported by the Appropriate Technology framework, were:

- adapt the project to the habits of the community and not the other way around;
- adapt the project to the local constructive culture;

- use local knowledge to design the solution;
- create a project tolerant of change and adaptation;
- create a closer relationship between the community and the project through collaborative work.

These ideas were structured into four major stages, as described below.

*7.3.1.1 The nature of the problem*

Data about the irrigation demand, the quality and availability of water sources, construction materials, service suppliers, distances for material transportation, cultivated vegetables, habits/period of cultivation in the year, crop area, inherited construction habits, means of transportation, and team daily costs were obtained from IBAP, the local community (Campada Maria) leaders, and the first mission survey on site. The team visited another collective garden with a similar but abandoned system, which reinforced the preliminary idea of searching for a more resilient and appropriate technology. Some initial agreements were made with the community regarding, for example, the task force (as a way to train for future maintenance), the feasibility of the implementation, and future reproductions of the project.

*7.3.1.2 Solution design*

A technically commonplace solution was designed, but with an approach that applied the principles of Appropriate Technology.

It consists of solar panels for energy generation, conditioned by means of an electronic converter; and the generated electricity is used to feed a motorised pump. This system avoids the use of batteries, replacing them with gravitational potential energy storage. The designed solution stores the pumped water in a reservoir placed on a masonry structure of stabilized adobe (soil, water, and cement), at the proper height for the irrigation of crops by the action of gravity through pipes. These decisions contributed to a reduction in the required investment, lower maintenance costs, and increased system life, as batteries are the first components to be replaced in photovoltaic generation systems. The same reasoning regarding the reduction of maintenance was applied to the choice of a construction solution.

Reliable data on the water volume available in the well, the demand for irrigation, granulometry and soil resistance, and the type, price, and availability of construction materials could not be obtained in advance, despite contact with IBAP technicians during the design process. Several uncertainties persisted, which generated multiple project versions. The initial plans had high safety margins with unviable budgets, and were revised to be simpler, with less risk (e.g., by changing the shape of the reservoir structure from triangular to rectangular).

This phase of the project design also included the learning time of the team, given the limited initial repertoire of technical solutions, as well as the managerial capacity for the pilot's circumstances and the skills necessary for its implementation.

To overcome these challenges and plan the construction process, a small-scale model was built (figure 7.4), reproducing the conditions of the pilot with the maximum fidelity: mass production of the structural blocks, the division into groups of

(a) (b) (c)

**Figure 7.4.** Scale model experimentation: design (a), adobe production (b) and construction (c).

people, the absence of electricity, the tools and equipment necessary, the construction interfaces with photovoltaic electrical generation, and the pumping and distribution system, in an attempt to predict possible problems and, in particular, their solutions.

There were other challenges:

- the absence of partners with knowledge of agronomy;
- the limitations of time dedicated to the project and financial resources;
- slow communication with technicians in Guinea-Bissau.

Thus, the project reached a certain point at which part of the solution was left 'open', to be completed at the implementation site.

*7.3.1.3 Pre-evaluation of the proposed solution*

The third stage, that of evaluation and training, consisted of the arrival of technicians from IBAP at the laboratories of the Federal University of Mato Grosso do Sul, the presentation of the proposed solution, evaluation, and project adjustments.

During the project pre-presentation meetings and mutual decision-making, Guinean technicians, predicting community members' opinions, showed mistrust about the solution adopted for the construction of the reservoir (stabilized adobe resistance and durability).

These issues were discussed and aligned among the team in order to anticipate problems in the implementation stage. A schedule, people in charge, and goals for the working groups were prepared based on the pilot.

The planning of the construction and assembly did not include labor expenses, as the intended products exceeded the budget.

*7.3.1.4 Implementation of the pilot*

In the first meeting with the community, shown in figure 7.5, the general definitions of the project were presented and the collaboration of groups was requested for each

**Figure 7.5.** Team and community design solution and planning agreement.

goal. Several community members expressed distrust in the use of adobe for a structural tower, alleging and reporting the lack of mechanical resistance and the solubility of traditional adobes (without any stabilizer).

Such uncertainty was stated at the beginning and lasted until the end of the project, but it was not sufficient to make the execution unfeasible, as the community's commitment to the project remained. This was the **first finding**: adobe is used by the locals, but it does not mean that it is a desired material, and it is an option only because of the lack of resources with which to buy industrial materials.

On each work front, the collaboration was partial but constant (figure 7.6) and, naturally, the assembly of the pumping system, which was a novelty, was more attractive than the construction activities with which they were more familiar. Ms. Fina, the community leader, introduced reliable information about the habits of men and women in the community and her confidence about the gradual collaboration that would emerge as the new project was included in the daily life of the community.

The 'open' parts of the project were gradually defined as the materials were decided/obtained. Adaptations and design changes occurred with the irrigation system (pipes, connections), the system layout, the support structure of the water reservoir (slab and closure above 5 m), and the cover. These decisions were made taking into consideration the community's frequent feedback.

Schedule delays and frequent problems inherent to the construction process happened (transportation, high costs, non-standardized materials) but all of them were able to be resolved at the local, regional or, at worst, national level.

Although the power generation and water pumping system can be quickly assembled and made available, it is part of the system that depends entirely on imported parts which is the weakness of the system. During implementation, this required the replacement of equipment and parts and subsequent maintenance.

(a) (b) (c)

**Figure 7.6.** Working groups for adobe production (a, b) and construction and assembly (c).

The 'shared doing' of the project revealed the organization of the community itself: women in group activities and men in heavier, isolated, and more prominent activities. This collaborative process had considerable weight in terms of acceptance, identification, and belonging to the project.

**This led to the second finding** regarding the less familiar part (solar photovoltaic generation and water pumping): even though this was difficult to implement and required maintenance during and after the finalized project, it was not technically questioned. The conventional and very well-known part of the solution, despite maintaining the original characteristics of adobe production and being based on scientific knowledge, was difficult to accept and generated controversy about its reliability. Even keeping the traditional production process, the small changes made and the new use (a taller building without a large roof), constituted an innovation and a bold step for the accumulated experience of the community.

The pilot was completed on the 44th day of the 45 dedicated to the implementation phase. A documentary was produced by another team (https://youtu.be/fe-UbCXBvsE), documenting the process.

### 7.3.2 Coffee bean solar dryer and solar-energy-based irrigation projects—Jamaica

This project presents the implementation of a solar water pumping system for coffee tree irrigation and a coffee bean solar dryer in the Blue Mountains area, in Jamaica.

In Jamaica, agriculture is one of the main economic activities; it uses the mountainous area of the interior of the island for different crops in an agroforestry system. In such places, access is difficult, limiting cultivation and harvesting technologies.

Coffee from the Blue Mountains region, which enjoys prestige worldwide, is produced in the context of artisanal, family-based production, with little technology and low financial investment, but under strict quality requirements, enabling the product to achieve high added value and high market prices, which do not match the amounts paid to the producers in the initial links of the supply chain.

This production, in addition to banana production, guarantees the livelihood of several families. However, Jamaican society still experiences social inequalities, from struggling small farmers to huge enterprises. Besides these stakeholders, there are middlemen who export coffee. In the mountain region, living conditions are extremely simple and access to infrastructure networks is limited, and the same conditions can be seen in the practice of agriculture, which is evidence of the lack of social and supply policies.

Since coffee is difficult to cultivate and also anticipating the effects of climate change on agriculture, the project aimed to provide more autonomy to small farmers and was developed by a partnership between the Fulbright Commission/USA (funding), the Universities of Technology, Jamaica, the Pontificia Universidad Católica del Ecuador, Ecuador, the Federal University of Mato Grosso do Sul, Brazil, Boston University, USA, the Universidad Nacional Autónoma de México, Mexico, and the coffee growers' association.

#### *7.3.2.1 The nature of the problem*

The first visit, involving several researchers from different areas, assessed the condition of coffee production in the region, identifying obstacles and prospective solution strategies in cooperation with local small farmers. Two issues were chosen to be the targets for design solutions:

- under normal conditions, coffee trees do not require irrigation, but in recent years, the supply of rainwater in the region has ceased to be regular, causing prolonged droughts. For this reason, irrigation has started to be considered by producers, to guarantee more reliable production. Such a system consists of a water source (a well or reservoir), photovoltaic energy generation without batteries, pumps and piping. Some parts of the system can be made by local producers to reduce costs, but most of them are unfeasible to produce, as they require high-technology industrialization;
- the coffee is sold right after harvest, still in the cherries, which reduces the bargaining power of the farmer, due to the product's perishability in its fresh state. For drying purposes, backyards are used (which require manpower to handle), or dryers, which can be solar. To make such dryers compatible with the producers' income, a low-cost solution that can be made by the farmers themselves was required. A temporary solution, consisting of a solar bamboo dryer with a very low construction quality, was assessed there by the researchers, and became a precedent in the search for a definitive solution.

Both of these demands aim to reduce coffee producers' vulnerability to climate change; in the case of the former by addressing water scarcity and productivity losses in the dry season and in the case of the latter by improving the quality of the coffee produced as well as reducing labor costs.

*7.3.2.2 Solution design*

Two products were designed; a photovoltaic battery-free water pumping system capable of irrigating 54 000 coffee trees over 60 acres on five coffee farms, and a bamboo solar dryer that allows the production of better quality dried coffee beans in a shorter time and which requires less labor to be built by the rural producers themselves, with or without executive guidance.

The goal was to meet most of the requirements of Appropriate Technology: to be low cost, easy to build, robust, and effective and to make the maximum use of locally available parts and materials.

The bamboo solar dryer was developed as a $2 \times 4$ m module (for 0.8 to 1.2 $m^3$ of coffee cherries), which can be extended by the addition of more modules. It consists of stabilized earth bags as a foundation, an elevated screened platform, a plastic cover, roll-up doors, and side openings to exhaust hot and humid air; the design is based on a Colombian reference (Oliveros Tascón *et al* 2008). The choice of materials and execution techniques sought to be:

- simple to make with the usual tools in rural areas;
- quick to make;
- made of easy-to-buy industrialized parts that can be found in small shops;
- not too bulky for transport;
- adaptable to different terrain slopes.

This dryer can cost 90% less than a conventional solar dryer, made with an iron structure and a plastic cover. Since the production is organic, the treatment of bamboo pieces places restrictions on the products that can be used. Although there are non-toxic treatment processes (using fire, for example), the project did not specify which one to use and focused on detailing how to avoid exposing parts to the weather and how to facilitate replacement in eventual exchanges.

The other product, the irrigation system, consists of a water dam, a surface pump, photovoltaic panels, an inverter, PVC piping, an elevated reservoir, and distribution by gravity for sprinkler irrigation. The panels are supported by foldable bamboo supports to adapt to different production and transport conditions (car or mule). The PV panels are the lightest part of the system and are is exposed to windstorms, so they were anchored to the ground by a base of rammed earth, which also increased durability.

To facilitate the future replication of these projects, online construction guides were made available.

*7.3.2.3 Pre-evaluation of the proposed solution*

Before the implementation of the pilots, prototypes for both systems were built and tested in a laboratory at the Federal University of Mato Grosso do Sul in Brazil (figures 7.7 and 7.8).

The three prototypes of the dryer and the five supports proved to be tolerant to a diversity of bamboo diameters and different levels of finishing. Despite the success of the tests, there was no participation of end users to analyze the solutions developed, due to the low budget. Likewise, agreements on how to carry out the transfer have

(a) (b) (c)

**Figure 7.7.** Solar dryer design (a), prototype construction (b), and endurance test (c).

(a) (b) (c)

**Figure 7.8.** PV panel support designs (a, b) and prototype endurance tests (c).

also been partially realized. The local partners were in charge of the choice of location in which to implement the pilots and dissemination to the community.

*7.3.2.4 Implementation of the pilots*

Along with the pilots, workshops were held for the purpose of technology transfer, so that the farmers would be able to reproduce the systems.

In contrast to the plan, and following a decision made by the coffee growers' association, each product was presented to different communities from different cities without previous agreement with the research team. In the solar dryer case, although the bamboo pieces were not pre-cut, selected, or dried, it could be built in three days, with the participation of six farmers, using simple tools and synthetic explanations, which demonstrated the self-explanatory capacity of the pilot.

The pilot was built on a collective site belonging to the association (figure 7.9), but even though it was built on private property, specialty coffees are vulnerable to theft, and the participants requested a design change to the way that the coffee is arranged for drying. Thus, a solution for removable trays was jointly created. That change in the design was made based on community feedback.

The dryer was evaluated by the farmers as being an appropriate product which solves the problem of labor demand, accelerates the drying process, even in a humid and very variable mountain climate, is easy to build, and generates ideas for subsequent adaptations and variations.

The irrigation system pilot uses a one-horsepower pump, six photovoltaic panels of 245 kWp each, and an inverter rated for 1.5 kW. The dam and water reservoir were left to the farmers; they were built from concrete-block masonry and positioned on a high piece of ground. The supports for the PV panels were quickly cut and assembled in one day, with the involvement of four people (figure 7.10). Despite the

(a) (b)

**Figure 7.9.** Training for (a), and assembly of, a three-module solar dryer pilot (b).

(a) (b)

**Figure 7.10.** PV panel support anchoring (a) and assembly (b).

fact that it was agile and inexpensive, those farmers considered that the cost of an iron or aluminum support would not be as impactful (10% to 15%), compared to the total cost of the water pumping system, a perception that could not be the same for other farmers. However, these same farmers perceived value in the solar dryer built with the other group of farmers, but given the scarce time for the transfer, a second dryer could not be built.

Each product was promised to a different group in a different city, overloading the already minimal team. In both cases, few people participated, because the community was not properly informed about the action by the farmers' association involved. Both products were very well accepted by the small farmers as adding high value, because they are effective solutions to the usual day-to-day problems faced. However, although the partnership could be justified technologically, it was clear that those involved in the execution of the pumping and irrigation system did not depend on the support of a project like this, as the labor was hired. As the project was developed for small farmers without resources to invest, it can be said that the pumping system did not reach the expected audience.

These results led to the **third finding** of the actions carried out: the products had value, but the Appropriate Technology had no value to one of the groups. This mismatch between the team's proposal and the public need was not caused by a design error, but by the deviation away from the target audience. These two problems could have been resolved by carrying out a pre-assessment with the community.

The future goal is to include a preservative treatment for bamboo that is easy and inexpensive, with low or zero toxicity, and accepted for organic production, in order to give the bamboo components a longer useful lifetime.

## 7.4 Project assessments based on theoretical approaches

Complex thinking, general systems theory, and concepts present in cybernetics are used to understand the relationships in collaborative processes. It is understood that processes based on participation and collaboration form a system in which actors from different universes create complex relationships among themselves, which can lead to different results (Nojimoto 2014), as these relationships are built over time and through learning from experience.

Complexity, according to Edgar Morin (2011), deals with the articulations between disciplinary fields that are dismembered by reductionist thinking, implying the recognition of a principle of incompleteness and uncertainty. Complex thinking is not thought based on imprecision, despite including it, but 'its purpose is to serve as a basis for building concepts and methodologies—tools for thought, reflection and action in the world—to articulate specialized knowledge' (Fiedler-Ferrara 2010).

In the case of the experiences previously reported in this chapter, it is clear that the proposed problems are of a complex nature, involving not only physical and cultural distance, when a team of researchers moves from its own professional performance system towards others, but that such problems also demand the capacity to solve very specific problems in the context of scarcity of resources of multiple orders, with very strong technological restrictions in the different locations where they operate.

Another principle addressed by Morin is that 'complexity is effectively the fabric of events, actions, interactions, retroactions, determinations, accidents, which constitute our phenomenal world' (Morin 2011, p 13). Theories involving systems and complexities assume that phenomena cannot be understood by a study of their parts in isolation, but by studying the relationships that compose them.

In this perspective, a participatory process that starts with the process of exchange between the actors involved can result in an emergent phenomenon. Morin defines emergencies as 'the phenomena that start from a system, which is an organized set of different parts, producing qualities that would not exist if the parts were isolated from each other'. These emergencies are desired in a complex system, which is open to improvisation, in order to enable new learning.

Observing the experience of the project implemented in Campada Maria in Guinea-Bissau, the emergency occurred, changing the expectation that the researchers had in relation to traditional construction techniques (adobe). Precisely because it is accepted know-how in the region, it was expected that there would be no questioning of the technique, but an understanding that it would be accepted and incorporated by the local community. But the fact is that the community actually embraced, without reservation, the unfamiliar technology, consisting of the PV panels, the inverter, and the water pump. On the other hand, for the familiar technology, they had reservations about the approach that the researchers initially proposed. Considering the systemic relationships between the actors involved, the 'open' project welcomed the unpredictability of the action, allowing an innovative result for the experience of the local community.

In the experience of building the dryer in Jamaica, researchers came across the information that specialty coffees are vulnerable to theft even though they are on private properties. Based on this unexpected fact, they had to think of a different solution for the dryer, departing from the prototypes produced, and the design of the trays was adapted to make them removable in an act of collective decision-making.

The conception of autopoiesis described in Maturana and Varela (1994) complements the understanding that systems can self-regulate and spontaneously self-manage in nature. Cellular processes are an example of this theory. However, what has been verified is that systems are not isolated from each other; on the contrary, the higher the level of complexity, the greater the level of interactions and possible connections for exchange, and consequently, the greater the connections that can lead to an imbalance between parties. From this perspective, we could say that there is no unrestricted autonomy, since an imbalance in one system will necessarily cause an imbalance in others.

In the participatory process, it seems that this condition of system self-reference is inherent to the process. This means that from the beginning, the participants should understand that they are part of a larger system. When this understanding is incorporated, community engagement tends to be greater.

The experience of building the irrigation system in Jamaica has shown that the lack of involvement of the right community resulted in a product that was only partially relevant to the community. The irrelevant part of the product was the one based on Appropriate Technology (bamboo-based PV support), although the other part was well accepted. This is largely due to the fact that the community was not properly selected.

Second-order cybernetics, as proposed by Heinz von Foerster, updates the notion of observing systems in which the observer himself is also observed. Observed systems could be classified as trivial machines, in which the inputs determine the outputs. However, when observation systems are living systems, they will never be trivial machines. In this way, the team of researchers also becomes an object of research, in the measurement of their actions, relationships, and propositions. In this sense, the production of the documentary in Campada Maria helps one to understand the entire process, from the difficulties faced to the decision-making, whether it is correct or not. In addition, the initial principles of cybernetics, consolidated by Norbert Wiener, propose devices for regulating systems, such as feedback loops, which in their cybernetic conception are strategic points in the process's metadesign, from which alternative paths are capable of re-establishing the systemic balance using a directional correction; after which, the overall result advances to a more effective status in the direction of the established goals.

To design a system in which more collaborative strategies of the bottom-up type can be implemented, is to design feedback and performance systems that consider self-regulation. This type of approach makes it easier to generate spaces for exchanges and transactions between participants, so that control and feedback mechanisms can facilitate such exchanges and seek certain levels of balance and performance.

Another contribution from the theory of complex systems is the interdependence between subsystems. We can consider from the two experiences that the subsystems are, for example, all the actors involved in the participatory process, each with its defined role. Researchers, interlocutors, the local community, financiers: each of these groups is a subsystem, a part, from which the whole is formed by interdependence. This means that the system is dynamic and larger than the sum of the parts.

In the participatory processes carried out in Guinea-Bissau and Jamaica, this interdependent relationship was established in the communication and exchange of information between the actors. If there is a decrease in the communicative and informational flow, it causes the end of a dynamic system that will prevent an important element of the process: the phenomenon of emergency. Maintaining information flow and allowing unpredictable situations to arise are essential conditions for understanding collaborative processes and achieving the desired results without major frustrations.

Thus, the more open the metadesign, the more the initiatives and solutions on the part of those who participate in the task force will be well-received and adapted to the system. The more there are predictions of points of possible unpredictability that are controlled by the design of actions, the less the unforeseen will be perceived as an error or a failure. In fact, mistakes will be seen as intermediate stages that will lead to success. In the same direction, the deterministic thinking that, as a rule, accompanies technical production in fields of knowledge such as architecture, urbanism, and engineering can synergize with certain levels of unpredictability and emergency, in the sense of what emerges as a solution from within the process itself, and the knowledge developed by all participants in the task force.

## 7.5 Lessons learned and final remarks

This chapter addressed the challenges, rewards and essential conditions for the development of projects based on technologies for social inclusion in vulnerable and remote communities. Each community posed a different challenge related to the distance of the community from the researcher's home town and/or country, difficulties in finding industrialized parts/components, budget constraints, limited execution time, low community involvement, etc. In this section, these challenges are detailed, the lessons learned are presented, and recommendations about how to deal with the challenges are given.

- Compensate for the distance between the community and the team's home town and/or country:

  In many cases, the projects are executed in communities very far from the team's base, or in close but unreachable areas, such as conflict zones. For instance, the two projects reported were implemented abroad. Such a reality requires careful planning, because the implementation of the project has to be done at once. Furthermore, since traveling requires time and money, the number of assessment trips is limited. In the presented cases, the projects had just one trip for problem assessment and community involvement and one trip for implementation.

With so few opportunities to interact with the community, the communication problems can be reduced by the use of more than one information source (e.g., more than one farmer, more than one technician). The idea is to accelerate data transmission and amplify the awareness of the problem (when there is no scientific data).

The challenge is to catch more than a momentary picture, because the data gathered for a problem can be related just to a fraction of the time spent together with the community.

Another suggestion would be, as far as possible, to insert digital mediation strategies, facilitating remote communication and bringing stakeholders closer together. For example, bring existing structures to the process, such as schools, associations, and city halls that have access to the Internet, to be incorporated into the process.

We have learned from the global pandemic COVID-19 context how to hold meetings remotely using numerous platforms and devices that have been developed and provided free of charge for the community in general.

Bringing stakeholders closer to online communication can generate more engagement and pre-knowledge about the local reality and complement the later physical presence. Nowadays there are simpler ways to share Internet access. For example, a cell phone with access can become an important hub for the distribution of information in the framework of this desired network.

- Simplify the industrialized parts/components of the projects:

  Although the projects were designed to be implemented based on Appropriate Technology, they may not use 100% of the low-tech materials available and may require some industrialized parts/components. In some countries, these items are imported, and therefore difficult to find and expensive. Such a reality may slow down the project implementation and increase its budget. Thus, it is highly recommended to make really easy designs, i.e. to reduce the variety of types of construction, assembly materials, and dimensions, reduce the diversity of tools needed, and propose solutions that are quick to build. Vulnerable communities have many other things to do.

  It is also suggested that as detailed as possible a mapping should be made describing the possibilities for purchasing input materials. For example, when it comes to construction materials, who are the suppliers that are close to the community? What kinds of material do they provide? Is there enough inventory? Since the action is very specific and with a budget limit, this prior mapping of suppliers could also be included in the process.

- Dealing with budget constraints:

  Invariably, on one side, projects for vulnerable communities have a limited budget. On the other side, the team tries to make more out of the available funds. That is possible, to some extent, because of the use of appropriate technology, which allows a very high return for the invested resource. However, it is important to realize that the 'do more with less' strategy makes the project management hard on the team, because each cent can make a difference and cause money to run short for the project's conclusion, leaving no margin of error for unexpected costs.

  A budget item that needs special attention is the logistics for the team to get to and stay in the community or nearby. When projects are implemented in remote communities far from the team base, a significant part of the budget is used for traveling costs, in some cases reaching more than 50%. It is clear that the projects using appropriate technology will always have a reduced cost for the products/pilots, and this will further highlight the disproportion of travel costs in relation to other expenses. If so, do not be alarmed, the path is right. And, for sure, larger projects are desirable because the overhead to total cost ratio is smaller.
- Plan to deal with limited execution time:

  In general, a project has a long total duration. The duration includes periods for planning, assessment, solution design, and execution (implementation) time, among others. In general, the total duration is not so short, but the time for implementation is. This happens because it has to fit into the team members' agendas and the travel cost increases with the duration. A limited execution time is hard on the team members and the community because the daily workload becomes too high, and it also leaves almost no room for errors or to deal with unforeseen problems. Furthermore, the distance from the team base to the community makes a second trip virtually impossible. That poses a challenge that has to be carefully addressed by the team members and the community during the planning phase. Many models and prototypes helped to resolve the details of the project and plan the implementation. Still, nothing can be guaranteed.

  However, it is possible to think of the metadesign in a way that contemplates the unpredictability. Alternative routes within the process should be provided to assist in decision-making at unexpected times. This preventive measure can, at least, establish more flexible margins in relation to the surprises that come from the local reality.

  A process that involves a community and a group of researchers from different countries is based on the principle of difference. Therefore, thinking about alternatives and anticipating unpredictability can help in this process.

  It is known in projects like the ones presented here that the cost of transportation is high and necessary. Therefore, in addition to the 'official' means of transport, it would be important to know if there are alternative routes and means already incorporated into the daily life of that place, which can help in unforeseen circumstances.

- Interact in order to achieve community involvement:

  A key factor for project success is the community level of involvement. The level of involvement depends on the community's organization, the previous experiences of the community in collaborative/cooperative projects, the perceived value of the project for the community, good community leaders that have a good understanding of the community work dynamic, strong and good community mobilization before, during, and after the project, appropriation of the project solution by the community before its implementation, etc.

  Find the real need of the community. If the project is 'noise', or if the goal of the project and the user's need are not the same, the project becomes a burden, an extra job, to be abandoned in the end. And the more remote the community, the greater the cultural constraints, e.g., the division of labor between men and women.

  Provide evidence to the community that local knowledge is an asset with high value to that and other communities. And be prepared, because the appropriation of the local culture for the proposition of any solution requires extra preparation on the part of the technical team to answer all the questions that will certainly arise.

  Good leaders and interaction with the community are the most crucial ones, while the others can be fixed during project execution, although this is not recommended.
- Find a really good local partner:

  The existence of a local partner who really knows the community and the site and who has good communication skills is crucial so that the project can define goals that will really help the community. Thus, the risk of developing a solution that the community does not need or want is reduced. A good partner also helps with the initial survey, which will focus on the potential of the site.

  As we previously recommended, prior communication with the local community is of the utmost importance. But when this is not possible, we can at least consider that the approach via the interlocutors is broader in the pre-configuration phase of the process. Knowing the local reality better through these interlocutors and increasing communication can facilitate researchers' knowledge of the local community, and thus help to predict unforeseen events as well.

  The notion of the systemic relationships that exist between the different players should always be considered in the basic assumptions of any collaborative design process.

## Acknowledgements

The authors gratefully acknowledge the financial support of the Fulbright Commission, the Brazilian National Council for Scientific and Technological Development (CNPq), the Institute of Biodiversity and Protected Areas (IBAP), and the Jamaica Coffee Growers Association (JCGA) for the projects reported in this chapter. They also would to publicly show their appreciation for the technical support from numerous Brazilian partners in the areas of structures, hydraulics, materials engineering, etc.

## Bibliography

Fiedler-Ferrara N 2010 O pensar complexo: construção de um novo paradigma *V!RUS* **3** http://www.nomads.usp.br/virus/virus03/review/layout.php?item=1&lang=pt

Hazeltine B and Bull C 1999 *Field Guide to Appropriate Technology* 1 (San Diego, CA: Academic Press)

Hoff H 2011 *Understanding the Nexus. Background paper for the Bonn2011 Conf.: The Water, Energy and Food Security Nexus* (Stockholm: Stockholm Environment Institute)

Maturana H and Varela F 1994 *De máquinas y seres vivos. Autopoiésis: la organización de lo vivo* 2nd edn (Santiago de Chile: Editorial Universitária)

Morin E 2011 *Introdução ao pensamento complexo* 4th edn (Porto Alegre: Sulina)

Nojimoto C 2014 Construíndo diálogos: complexidade e emergência em processos de design *PhD Thesis* Universidade de São Paulo, São Carlos

Oliveros Tascón C E, Ramírez Gomes C A, Sanz Uribe J R and Peñuela Martínez A E 2008 *Secador Parabólico Mejorado 376* National Coffee Research Center - Cenicafé, Colombia Avances técnicos Cenicafé www.cenicafe.org/es/index.php/nuestras_publicaciones/avances_tecnicos/avance_tecnico_0376

Patnaik J and Bhowmick B 2019 Revisiting appropriate technology with changing socio-technical landscape in emerging countries *Technol. Soc.* **57** 8–19

United Nations Development Programme 2020 Human Development Reports - Guinea-Bissau http://hdr.undp.org/en/countries/profiles/GNB

**IOP** Publishing

Sustainable Urban Development

Topics, trends and solutions

**Luís Bragança, Cristina Engel de Alvarez and Luisa F Cabeza**

# Chapter 8

# Heritage conservation and sustainability

**Maria Matilde Villegas, Vladimir Fernando Stello and Ricardo Rodrigues**

Urban transformations and changes in heritage interpretation trigger new challenges that, in turn, force a rethinking of the management of historic cities. On the one hand, heritage conservation faces the consequences of urban growth, the emergence of new centralities, the loss of resident population, and, in many cases, urban emptiness, with the consequent environmental degradation and loss of vitality. On the other hand, the concept of heritage has evolved, and considers not only the monumental landscape, but also the ensembles, sites, traditions and, more recently, cultural landscapes. In this new scenario, the sustainability of heritage, particularly in historic centers, requires appropriate approaches to different objectives and specificities. In this chapter, we argue, with examples, for the attractiveness of historic centers as emblematic, symbolic spaces, but also as spaces that can continue to be inhabited and enjoyed; it is necessary to coordinate actions from an integral and sustainable perspective, that is to say, addressing social, educational, heritage, urban, economic, and environmental issues.

## 8.1 Sustainable development

The term *sustainability* is used in different discourses by different sectors of society, according to their own interests. The concept of sustainable development arose from the perception of the problem of the disorderly growth of civilization. Accordingly, both its theorization and its practical application must be dynamic, following the global development process.

The sustainable development concept is relatively new. References to sustainability appear in the 19th century, referring primarily to its biological sense, according to which, a species should have the ability to maintain itself in a given environment without generating a degradation of resources that threatens its existence in the future (Moreira 2004).

Until the Industrial Revolution, development was measured by the expansion of intellectual and material capacities. Since then, it has been evaluated using the accumulation of goods that resulted from the increase in production and

doi:10.1088/978-0-7503-3971-1ch8

consumption, which began to regulate social, political, and everyday actions. This vision results from a belief in the unlimited availability of natural resources, ratified by precepts such as: what matters is the individual, mankind can and must have full control over Nature, economic determinism is common sense, and technology will fix whatever is necessary (Bateson 1987).

In the mid-twentieth century, this posture began to be questioned, with the publication of 'The Limits to Growth' (Meadows *et al* 1972), disseminating the discussions of the so-called 'Club of Rome' and the mathematical projections of the failure of the development model adopted until then. International organizations promoted meetings and recommendations, such as the World Conference on the Human Environment (1972), the 'Our Common Future' Report, also known as the Brundtland Report (1987), the Conference on Environment and Development—Rio 92 (UN), the World Conference on Sustainable Development—Rio+10, and the Johannesburg Summit (2002), in order to discuss and modify this pattern. Of these recommendations, the best known is the Brundtland Report, which defined sustainable development as follows: *sustainable development is development that meets the needs of the present without compromising the ability of future generations to meet their own needs.*

However, this concept also had to be improved, mainly because it involved a more quantitative than qualitative notion, reinforcing the capitalist logic of development. In this context, several contributions have been taking place in order to expand this definition. Since the end of the last century, the International Council for Local Environmental Initiatives has understood that, to be sustainable, development should provide the entire community with environmental, social, and economic services, without putting natural and social resources at risk. Sachs (2007) in his definition of eco-development (1973) proposed that the ecological, spatial, and cultural dimensions be added to the social and economic dimensions.

The Organization for Economic Co-operation and Development (OECD) addresses the issue of the need to focus on sustainable development by considering the quantity and quality of growth, understanding the concept of quantity as growth and quality as development. Sustainable development is beginning to assert itself as a necessary alternative to dominant behaviors in which the consumption of objects, people, and places is associated with the idea of happiness (Zanirato 2016).

In this context, it can be said that sustainability depends on an intellectual revolution, a cultural, scientific, and social metamorphosis that implies treating the economy, culture, and environment without hierarchies. They are parts of an interacting whole and cannot be treated separately. This notion breaks with a long tradition of prioritizing the economy over the environment and the preservation of cultural heritage. For this development to be fully achieved, the dimension of culture must be incorporated. A more harmonious and fair development must encompass the interests of society with regard to material and non-material cultural heritage.

## 8.2 Preservation of cultural heritage

Cultural heritage is a fundamentally important part of forming community. In its Latin origin, the word heritage—*patrimonium*—meant the set of goods that a person owns. When it is interconnected with the word cultural, which in its origin—*culturae*—means 'action to treat', 'cultivate' or 'cultivate the mind and knowledge', it then encompasses a much broader context, that of culture as the people's heritage. It is related to the people's own identity, the characteristic elements that differentiate it from others.

In the broadest sense of the word heritage,[1] every built environment would fit its definition. However, not all interventions are invested with meaning by society. In other words, not all of them are recognized as cultural heritage. According to Poulot (2008, p 36) such recognition depends on an attribution of value:

> Briefly stated, the notion of heritage involves, first, a set of possessions that must be identified as transferable; then, a human group, a society, capable or susceptible of recognizing them as its own, of demonstrating its coherence and organizing its reception; and, finally, a set of values, political in the most general sense of the term, which allows articulating the legacies of the past while waiting for, or shaping a future, with the objective of promoting some mutations and, at the same time, affirming continuity.

The attribution of value based on the selection of certain elements is essential for the identity of citizens in a given place:

> As links in a chain linking the present to the past, transforming the past into a dimension of the present, the material cultural heritage of a social group is one of the fundamental elements for the construction and consolidation of collective identities. It is also the support of a plural memory that surpasses time and space. (Benchetrit 2008, p 20)

In this context, a group of people representing society—technicians, politicians, organizations, associations, etc.—establishes a selection process for both what it will protect, for the future, as a material reference, as well as replace, by not giving it any special meaning, or as a way of erasing certain memories. In addition to architectural and natural elements, this selection includes a whole range of interventions carried out by Man in the environment where he lives, as well as traditions and customs, in a close relationship in which the urban environment is the product.

> Preserving the urban environmental heritage is, as can be seen, much more than simply toppling certain buildings or complexes: it is, rather, preserving the balance of the landscape, always thinking as interrelated to the infrastructure, the lot, the building, the urban language, uses, historical profile and the natural landscape itself. (Castriota 2009, p 89)

The official concern for safeguarding heritage in Europe was consolidated from the second half of the 19th century, while in Latin America this concern occurred

[1] Considering the accumulation of investment in the form of labor and financial capital.

later, from the first decades of the 20th century. Since then, these concepts have evolved from a set of monuments representing selected national oligarchies, to myths related to the formation of nations, to material and immaterial goods, reaffirming the identities that characterize a society, or those that are part of it, and that contribute to the formation of an idea of nationality, and its identity.

In Portugal, the problem became more notorious after 1834, the year in which the religious orders were extinguished. The difficulty of conserving the built heritage gradually became evident, a problem that would be aggravated in 1911 after the law on the separation of state and church came into force. The first monument to undergo restoration intervention in Portugal, still in the 19th century, was the Monastery of Batalha, whose works were financed by the King's consort. The idea of 'monument' gained strength, amid demolition movements and the growth of cities. Often, the actions related to the valorization of the monument imply large-scale demolition processes and urban transformations, highlighting the 'monument', a fact that today is recognized as criticizable, because it takes the building out of the context that gave it a spatial sense. However, the idea of 'monument', as a relevant element in urban space began to assume importance much earlier. According to documentation from the 15th century, the stones of ancient and significant buildings could only be moved with royal authorization. This same concern can be seen in relation to the primitive church of St. Tiago, in Guimarães, which collapsed in 1607. It was then replaced by a small chapel, apparently 'not without murmuring from those who thought it had more credit to preserve the ancient ruins than to reduce them to such smallness' (Ferrão 1999).

Nevertheless, the 20th century is the period in which the greatest and most evident actions, with a sense of heritage defense, have been carried out. In Portugal, the creation of the Directorate for the National Buildings and Monuments (DGEMN) constitutes a benchmark in the sense of enforcing the ideological formula of the *Estado Novo*: 'God, Fatherland, and Family' (Neto 2001, p 140), preserving and intervening in those historic monuments with which the political regime identified itself, or sought to highlight, emphasizing the idea of the *Portuguese Fatherland.* In the second half of the century, especially after the 1974 Revolution, the social dimension and the problems of housing and sanitation gained prominence. In older cities in particular, their 'historic' nuclei, often densely populated, presented problems of the physical degradation of buildings and unqualified public spaces, which generated the phenomenon of segregation of the resident population, who were generally poor and living with illegal practices, namely drug trafficking, crime, and prostitution. The seriousness of the situation, in the Portuguese case, led to the creation of several operations, such as the SAAL, a specialized technical body, which aimed to support, through municipal councils, initiatives amongst populations living in bad housing, collaborating in the transformation of their own neighborhoods, investing their own latent and, eventually, monetary resources. Also, a few years later, in the mid-eighties, the Urban Rehabilitation Program (PRU) was implemented and Local Technical Offices (GTLs) were created in association with the municipal councils.

In Latin America, some movements are considered landmarks in the search for heritage preservation, such as the one that occurred in Brazil in 1920, known as the Week of Modern Art, which boosted the National Historic and Artistic Heritage Service—SPHAN and instituted listing structures as a legal instrument of protection.

In the late 1980s and early 21st century, in Latin American countries, the concept of heritage was expanded. Not only were monuments and colonial ensembles still important, but also a wide range of buildings and historical ensembles, produced at different times, with different typological characteristics, in which the variety of heritage is important (figure 8.1). Subsequently, assets of an immaterial nature were included in the preservation categories, as well as cultural landscapes.

At the beginning of the 20th century, as advocated by the 1933 Charter of Athens, the principle of limiting preservation to isolated monuments as exceptional representatives, testimonies of an era surpassed, was taken to the extreme by modernist architects, in their eagerness to redesign the existing city. Allied to this were the real-estate interests that preached the verticalization of urban centers, which practically subjected the entire city to demolition. In this period, the listing of isolated buildings was a possible alternative for the preservation of the built heritage.

From the 1970s onwards, as an alternative to the listing of isolated specimens, the principle of preservation by zones was established. In an attempt to integrate the urban planning process and its zoning initiatives, it was believed that the control of growth and urbanization would include the preservation of historic centers and other areas of interest, which should be protected by restrictive zoning. Historic centers and zones should play the parts of specific solutions in the urban framework, reinforcing the functionalist and schematic principles of zoning. However, this rigidity of legal restrictions on land use and occupation led to the freezing of historic neighborhoods, resulting in economic stagnation and environmental deterioration.

In recent decades, to overcome the problems created by restrictive zoning in historic areas, a new level began to be sought by preservation policies in many countries. It was noticed that the stagnation of protected regions did not contribute to their preservation. Thus, the view of historic buildings as monuments or museum pieces was replaced by the idea of the city as a cultural complex formed by the overlapping of different periods.

Taking into account the economic, social, cultural, and symbolic capital of the central areas of cities, the urban planners proposed, in contrast to the model of the Charter of Athens, new parameters for urban planning. The propositions can be organized into two types: one-off, in which a single work takes on the role of revitalizing the area in question; and the other, with a broader scope of intervention, in which a region is considered, aiming at social inclusion, economic revitalization, and the preservation of cultural heritage. The most obvious examples of the first type are the Sydney Opera House (Australia), the Guggenheim Museum in Bilbao

**Figure 8.1.** Panoramic view of the historical centre of Laguna, seen from Santo António lagoon (from the collection of Marco A. de Souza 2009).

**Figure 8.2.** Panoramic view of the historical centre of Guimarães (Arquivo CMG).

(Spain), the Museum of Tomorrow in Rio de Janeiro (Brazil), and the Convention Center in Cartagena de Indias (Colombia). The second type refers to examples such as the Historical Centers of Bologna (Italy) and Guimarães (Portugal) (figure 8.2) and the Cultural Corridor in Rio de Janeiro (Brazil). These interventions, in addition to the positive results in relation to cultural heritage and the revitalization of its surroundings and environments, served as a counterpoint to the renovation policy that was responsible for the demolition of parts of many of the historic city centers around the world, erasing a significant part of the history and identity of important cities and their cultures.

In Brazil, the advent of the document called the City Statute, in 2001, officially expresses the concern for the inclusion of the preservation of cultural heritage as an important element in city planning. It should be seen not as an isolated or detached element of the city, but treated as an integral part of the general context of the city.

> Adopt an unitary procedure, aiming at the improvement of the urban environment as a whole, not treating the so-called historical areas, and the other spaces that make up the city, unequally. Thus, it is important to think together, and based on the same quality criteria, on spaces shaped by history, to be protected, as in new spaces or recoverable spaces, to be structured. In both cases, the infrastructure, the lot, the building, the urban language, the uses, the historical profile and the natural landscape must always be considered. (Castriota 2009, p 90)

Thus, cultural heritage is considered a fundamental element for the recovery of identity and citizenship.

> Knowing, perceiving and feeling the city's history is to become a citizen, it is to develop the feeling of identity with the environment as a part of it. Cities are

> growing, transforming, and the population is losing the references of its urban evolution. The new generations, not having the experiences of the past, cannot perceive the values left by their ancestors. It is necessary to remember, and to create mechanisms, in order to quickly access the past and our history. A society aware of its past is better able to move coherently into the future. Any study of the city, whether specific or in the context of urban planning, will not escape from seeking, in the recent or more distant past, the evaluation of its development process. (Souza and Muller 2007, p 9)

Cultural heritage results from a dialectic between human beings and their environment, between the community and its territory, between the present and the past, and between these and the future. In the updated view of the concept, it came to be seen as a fundamental resource for sustainable development, in which its current use should not be compromised for future generations.

Therefore, it is necessary to reflect on the concepts that guide the practice of preservation from the perspective of a balance between practice and representation, materiality and immateriality, in addition to the different values given to heritage by different users. Such values cannot be considered in isolation, because the different elements that make up the cultural heritage cannot be inserted, one by one, into small watertight niches, but form a set of factors that make up society. In this sense, it can be said that the preservation of cultural heritage is a political act, not in the partisan sense, but in the sense of the Greek origin, of the *polis*, administered by the citizens as a community.

From the above, we understand the importance that the cultural property has for its social function, as a bearer of a message from the past, whose meaning is reinforced through its enjoyment in the present. For this, interventions and strategies are needed to make it accessible to the population and support the recovery of its potential.

## 8.3 The perspective of sustainability in heritage conservation

The inclusion of the concepts of sustainable development and sustainability in heritage preservation is relatively recent, appearing from the 1960s onwards in international documents. The study 'The "archeology" of sustainability in the conception of cultural heritage' (Cantar *et al* 2021) analyzed the evolution of the link between the concept of sustainable development and heritage preservation in six stages, considering a starting point of 1964.

### 8.3.1 First stage: 1964 to 1972

In the 1960s, there was no clear or normative articulation between heritage and sustainability. However, experts and researchers already shared ideas and criticisms of the development model used up to that time. Conservation for the preservation and safeguarding of cultural and natural heritage was seen as a primordial measure, considering the risks inherent to accelerated urban transformation. The first movement of international importance that focused on environmental issues was the First

International Conference on the Biosphere, organized by UNESCO (Paris 1968), where the development model applied up until that year was questioned, as well as the rational use and conservation of natural resources.

In the area of cultural heritage preservation, the 'International Charter on the Conservation and Restoration of Monuments and Sites' in 1964, commonly known as the Venice Charter, marked the beginning of the creation of specific guidelines for preservation actions in the built environment, highlighting the need to preserve monuments while keeping them connected to their surroundings. However, there is still no mention of, or correlation between, the conservation of monuments and sites and their socioeconomic or environmental contexts.

### 8.3.2 Second stage: 1972 to 1990

At the United Nations Conference on the Human Environment held in Stockholm (Sweden 1972), 26 principles were identified that constituted an Environmental Manifesto and laid the foundations for the United Nations environmental agenda. This period was marked by the consolidation of the environmental issue on the international agenda, which linked the concepts of sustainability and development that were made official in the Brundtland Report (*Our Common Future* 1987), in which it is recognized that the progress of society and of the world depends on actions that aim to preserve the planet. In this context, the concept of the sustainability of natural and cultural resources is associated with economic aspects, considering the need for a balance between parties for an effective and long-term policy. To do so, it was necessary to coordinate measures and promote development supported by comprehensive planning.

### 8.3.3 Third stage: 1990 to 1999

In this period, the importance of community participation is highlighted, as well as the emphasis on the importance of valuing local identities and indigenous peoples. One of the main documents was an update of Burra's Charter (1979, updated in 1981, 1988, and 1999), which surpassed the classic view of conservation that took into account mainly historical and aesthetic values, seeking to value social and spiritual values in the same way, calling this 'cultural significance'. Heritage preservation was also incorporated into urban planning policies and these new dimensions were linked to the concept of sustainability. The management of natural and cultural resources began to be seen as an instrument of territorial planning, and the notion of sustainable development was widespread and popularized. In practice, it was officially used through the establishment of public policies by several countries, at different levels of development, and by the community in general, through specific actions. At the United Nations Conference on Environment and Development, called 'La Cumbre de la Tierra', or simply Rio 92, global agreements were directed towards the establishment of goals for the next century. The main objectives of the so-called Agenda 21 were defined, based on a global strategy aimed at achieving sustainable development supported by global cooperation. Twenty-seven principles were established to guide the actions of member states in order to ensure the progress and well-being of humanity; the creation of legally binding commitments between different governments was an additional challenge (Cantar *et al* 2021).

### 8.3.4 Fourth stage: 1999 to 2011

The UN World Conference on Sustainable Development (Río+10, Johannesburg Summit 2002) marked the fourth stage. Its main objective was to find alternatives for the application of Agenda 21 in Rio de Janeiro, fundamentally through community participation. This period represents a central milestone in the relationship between sustainability and heritage, which becomes a theme present in the documents produced by ICOMOS and UNESCO. It was a period of great production of international documents that emphasized the importance of a more integral preservation, including and considering both the material and the immaterial, the diversity, and the cultural landscape. Among the most relevant documents are the Kimberley Declaration on the Intangible Heritage of Sites and Monuments (2003), the Convention for the Safeguarding of the Intangible Cultural Heritage (2003), the Convention on the Protection and Promotion of the Diversity of Cultural Expressions (2005), and the Quebec Declaration on the Preservation of the Spirit of Place (2008).

In apparently contradictory movements, the non-inclusion of culture as one of the Millennium Development Goals (MDGs) defined in 2000, through the Millennium Declaration, adopted by the 191 member states, was questioned. In 2007, the UN's Intergovernmental Panel on Climate Change—IPCC (created in 1988) was awarded the Nobel Peace Prize, as a result of studies related to sustainability, thus moving to a new dimension of perception and acceptance by society.

### 8.3.5 Fifth stage: 2011 to 2014

In 2013, the 'Hangzhou Declaration' placed culture at the center of sustainable development policies, reinforced by the 'Florence Declaration' on *Heritage and Landscape as Human Values* (2014) which addressed culture, creativity, and sustainable development. As Cantar *et al* (2021) state, it is evident that:

> (…) the process of strengthening and expanding sustainability in the heritage concept. Environmental problems of international scope and their links with the safeguarding of cultural assets were recognized. Furthermore, emphasis was placed on the role of culture as an engine of sustainable development.

At the conference 'World Heritage and Sustainable Development: the Role of Local Communities', held in Kyoto, in 2012, which resulted in the document known as 'The Kyoto Vision', it was recognized that the concept of heritage is fundamental for sustainable development, as it results from the interaction between communities and their surroundings. This confirmed the idea that it would only be possible to achieve this goal if the relations between the population and its heritage were based on respect for cultural and biological diversity, including material and immaterial aspects, and oriented towards sustainable development.

### 8.3.6 Sixth stage: post-2015

At the Sustainable Development Summit (2015), UN member states approved the 2030 Agenda, establishing 17 Sustainable Development Goals (SDGs). In the replacement of the MDGs (2001) by SDGs, the most significant difference was the inclusion of transverse cultural cooperation and a deeper focus on fully rethought sustainability goals. This means that culture should work as an engine and facilitator of sustainable development. Although none of the 17 SDGs focuses exclusively on culture or cultural heritage, there are several references to the theme, such as targets 2.3 and 14.b that relate production and sustainability to traditional communities, in target 4.7 for cultural diversity, in 8.9 and 12.b which promote sustainable tourism based on local culture and products, and more specifically in SDG 11 'Making cities and human settlements inclusive, safe, resilient and sustainable' in its goal 4, when it states that it should 'strengthen efforts to protect and safeguard the world's cultural and natural heritage'.

Reinforcing this trend, in 2016, UNESCO organized the report entitled 'Culture: Urban Future: Global Report on culture for sustainable urban development', proposing concrete guidelines to support decision-making and the management of cities, based on culture as an engine of development. This theme also appears in the Report on 'Culture in the Implementation of the 2030 Agenda' (2019), which ratifies the link between sustainability and heritage. With the incorporation of culture in the framework of the SDGs, policies, objectives, and goals were placed within the scope of an international agenda.

## 8.4 Conservation as a principle for sustainability

At the end of the last century, it was found that 40% to 70% of urban solid waste collected in large and medium-sized Brazilian cities came from construction, and it was estimated that even after more than two decades, this reality would still remain (Pinto 1999). The author also states that, according to the reference estimates used in his research, the 'construction waste generation rate' could be defined at around 150 kilograms per constructed square meter. Regarding the topic, a report prepared for the National Trust for Historic Preservation (2011, p VIII) mentions that:

> This study finds that it takes 10 to 80 years for a new building that is 30 percent more efficient than an average-performing existing building to overcome, through efficient operations, the negative climate change impacts related to the construction process.

Such data highlight the waste of raw materials used to satisfy the current needs of housing and services, which causes large environmental impacts. Civil construction represents the human activity with the greatest impact on the environment, but it also promotes influential socioeconomic aspects in improving the quality of life. That is why it is such an important goal in the implementation of each country's sustainable development plans.

The search for more efficient solutions in the use of natural resources, with less environmental impact and higher social justice, must accompany the improvement in the quality of life, as a new (sustainable) development model. This process must consider the different areas, namely the civil construction sector and, more specifically, the sector of conservation and restoration of buildings, public spaces, and historic centers. The civil construction sector is facing new challenges in the search for sustainable strategies that must address the different stages from project design, through the construction, operation, and maintenance stages and, if possible, consider the period up to the end of the useful life or *retrofit* of the edifice.

Norman Foster (2003, pp 11–12), at the beginning of this century, stated that:

> If sustainability is to be more than a fleeting fashion, architects in the future must ask themselves some very basic questions. For example, why do we still insist on using green-field sites when we could build on reclaimed land in our cities? **Why do we demolish buildings that could easily be put to new uses?** Why do we rely so heavily upon artificial lighting when we can design buildings that are filled with daylight? And why do we continue to rely upon wasteful air-conditioning systems in locations where we can simply open a window? [Bold added by the authors.]

Restoration, like the construction of any building, can be divided into several stages that include, among others: a survey of the existing building, the analysis of pathologies, the development of the intervention or restoration project, the execution of the work, and the use of effective maintenance procedures. Generally, the last step would be the end of the building's useful life, but what is aimed at with restoration processes, or even retrofit, is precisely to make it sustainable by prolonging its useful life.

The insertion of strategies that aim for sustainability into projects/designs for the conservation of a historic center, or the restoration of a building, must be a priority and be present at all stages related to the useful life of the built environment. The design of the conservation or restoration project, in particular, must seek creative mechanisms consistent with preservation practices as well as with the concept of sustainability. When present in all phases of the preservation process, sustainability has a strong capacity to induce cultural, social, and environmental change in a community.

Faced with the challenges of the sector, there is a need to review the concepts, methods, and forms of action of professionals linked to the preservation and restoration of the built environment, and this topic is widely discussed in international events related to the topic. This process must consider the educational component, namely the insertion of these themes into the educational formation of urbanists, architects, engineers, and all professionals who work at some level on the transformation of the territory. Planning, in the sense of programming investments for the development of the different territories, must, as a rule, require a previous impact assessment that may suggest ways of considering retrofit, and reserve profound transformations for a worst-case scenario that must be fully justified.

In the current situation, the use of natural resources in the preservation of the integrity and authenticity of a building is problematic, as the original materials are scarce or non-existent. But, what reflection should be made considering the comparison between the environmental impact resulting from the reuse of these materials in relation to new buildings? Or, alternatively, will it be possible to replace certain materials, innovating beyond traditional values and reuse and including aspects related to sustainability (including building use)?

In the light of sustainability, it is essential to incorporate the cultural component into the 'architectural design', which, in the case of buildings of heritage interest and historic centers, is the most important renewable resource in conservation actions. The recovery of knowledge, the permanence of construction techniques, the preservation of traditions, the training of artisans, and community appropriation are elements that must be integrated into a vision of sustainable preservation so that there can, in fact, be an integral preservation program.

A well-designed restoration project must understand the aspects inherent to its specific characteristics, which is a key point in making it a sustainable project. Good conservation practices are, in essence, sustainable, as their performance can be improved by the use of traditional construction methods and the principle of minimal intervention. Many techniques used when there was no thermal conditioning technology from an electrical device, such as thicker walls, taller, and often with small openings, help in thermal comfort without the use of energy-dependent mechanisms. The historic buildings, especially the larger ones, were designed with architectural elements that allow more natural light and ventilation, with high ceilings, large windows, glazed doors, large balconies and roofs that provide shade, managing the accumulation of direct solar radiation, which is mainly of great importance in buildings located in tropical countries. Historic buildings used traditional materials and technologies of local origin (figures 8.3 and 8.4), such as earth, stone, lime mortar, wood, among others. When these buildings are restored,

**Figure 8.3.** Intervention at the 'House of Rua Nova', Guimarães, Portugal (Arquivo GTL/CMG 1985).

**Figure 8.4.** St. Benedict's Church ruins, Maragogi, Alagoas (from the collection of Maria Matilde Villegas 2013).

logically, the same materials should be sought or, if they must be replaced, materials should be used that have similar or compatible physical and mechanical characteristics. This means that, in restoration practice, reused materials are often used, which, in principle, represents a sustainable approach. These same principles can be considered when it comes to thinking about new projects that are part of sustainable development.

Another advantage found in restoration procedures refers to the concept of energy incorporated in these buildings, since the reuse of buildings tends to strongly reduce carbon emissions from new construction. The same applies in relation to the reduction in the generation of construction waste, as many of the materials used, such as lime mortar, wood, and brick, have lower incorporated energies than those of modern materials such as glass, aluminum, and steel.

It should also be noted that these buildings tend to respond more appropriately to moisture problems, since many of the materials used, such as wood and lime, allow the building to *breathe* and thus help to control excess moisture, which, in addition to causing discomfort to users, can also lead to structural problems. In this sense, there is a consensus among the experts that seeking conservation actions for historic buildings, in order to keep them in good condition and in use, makes them sustainable. Although the focus here is on historic buildings, in fact, these principles apply generically.

Thus, one can see how the theme of sustainability is gaining ground, with an emphasis on buildings that are described as *heritage* buildings within the context of architecture and construction as a whole. But some doubts still prevail, especially in terms of how to make an objective assessment of these buildings. The evaluation systems used in modern constructions of sustainable architecture would not work for this type of building, as quantitative measures using a single criterion cannot be applied, and aspects related to the relevance of the building to society and its affective bonds can only be measured subjectively.

## 8.5 Some additional considerations

The field of built heritage preservation is in a complex situation in terms of how it acts, as it faces the challenge of recovering structures that were designed and built in social, economic, and environmental circumstances that are very different from the current ones. The same applies to materials, their quality, working methods, and the availability of material and human resources, which differs substantially today. Urban codes, safety, and thermal and acoustic standards are also a potential obstacle to the maintenance of the physical and structural characteristics of old buildings.

Historical centers are the areas where some of the most recognizable signs of societies are manifested and are characterized by containing the greatest density of goods linked to their history. However, within the framework of the intense transformations that cities are currently experiencing, these spaces have been subjected to various processes of change and reinvention of their historical landscape. In this new scenario, the sustainability of heritage, particularly in historic centers, acquires a specificity and demands strategies that have not been thought of before.

No one would question the importance of the preservation of monumental elements such as the pyramids of Teotihuacán in Mexico or the ruins of San Miguel Arcanjo in Brazil, so that future generations can enjoy them. But in the case of the preservation of urban cultural heritage, which includes material and immaterial goods with historical value that are present in a city, the relationship with sustainable development is even more direct and urgent, as these goods are more numerous, of national and local value, and experience intense daily use by the population. However, the urban cultural heritage of small towns is more anonymous and the importance of its preservation is more difficult to understand and promote. In this sense, it is important to remember that the preservation and good use of these spaces can provide:

- **environmental sustainability**, by reusing existing infrastructure and networks to respond to the demands that have to be met for the city to function. In particular, the preservation of historic centers can help to promote the compact city, given the existence of nearby services, and allows more people to live in spaces that are more connected to public transport. The old neighborhoods can, therefore, constitute a space of opportunities for the sustainable development of a city as a whole. As mentioned before, these processes and consciousness must be developed throughout the different stages of the educational process, in particular, those regarding the planning and building professionals. Raising awareness of the impact that each decision implies for the broader scenario is mandatory, and each decision maker—at all levels—must understand its importance at the ecological, environmental, economic, and social levels. This seem obvious in the heritage conservation field, but must be considered for the whole territory as a cultural footprint.

- **social sustainability**, considering that consolidated urban centers have a traditional social and cultural fabric. A strong cultural tradition is not an impediment to sustainable development. On the contrary, it is a character or a condition common to a place, with the political potential of being able to unite a community to work towards its demands. The recovery or maintenance of a place values the appropriation of its identity, whether by people or by their ways of life, which has developed in this common space.
- **energy, financial and economic sustainability**, since preservation itself represents a great financial saving. Since it is based on the assumption of taking advantage of existing resources, such as buildings, it therefore follows the principle of sustainability. Thus, profitability and growth can be increased through a more efficient use of local resources, including labor, materials, water, and energy, as it has a local productive tradition, with a high flow of people and complementary services. In general, the energy consumed to replace old buildings with new ones, even those that are more energy efficient, will not surpass the benefits that arise from the conservation of historic buildings, not to mention the social and cultural benefits. Therefore, preserving buildings of heritage interest as well as the historic sectors of cities, reusing spaces, valuing their environmental characteristics, and taking advantage of their incorporated energy (and avoiding their waste), makes sustainable development a viable path.

The traditional knowledge, with its materials and ways of doing and performing, is the articulating element between nature and society, between the past, the present, and the future. This patrimony can help in the search for balance in relation to the needs of today's society without, therefore, putting the patrimony at risk for future generations. The management of this heritage must be done in a respectful way; it is essential to consciously and efficiently take advantage of the cultural legacy as a factor for the incorporation of communities and their social dynamics, wherein the main premise is the sustainable conservation of the natural and cultural heritage.

This process is part of the integrated conservation of cultural heritage with the communities that hold it, being necessary to recognize the importance of vernacular knowledge and making it known, used, and replicated, for its transcendence and the improvement of the quality of life of present and future generations. The preservation of a community's heritage is only important insofar as it allows it to develop socio-culturally and in balance with Nature. Heritage conservation may be especially useful in the future if retrofit actions become more common each day more common. The knowledge and experience that may be learned from heritage conservation encompasses not only the material aspects, but also, and in particular, a programmatic, methodological shift. The word conservation must perhaps be reconsidered, as it still seems to generate a reaction from those who look at it as an obstacle to development. Continuous improvement and change is needed and natural. But concerning sustainability, this is one of the outcomes urgently needed: to clarify and overcome the (apparent) contradiction between development and conservation.

## Bibliography

Bateson G 1987 *Steps to an Ecology of Mind: Collected Essays in Anthropology, Psychiatry, Evolution, and Epistemology* (Northvale: Jason Aronson)

Benchetrit S F 2008 *Preservar por quê? Preservar para quem? Um olhar contemporâneo sobre a preservação do patrimônio cultural material* Museu Histórico Nacional, Rio de Janeiro pp 20–5 ed. C S R de Carvalho *et al*

Cantar N M, Endere M L and Zuleica M L 2021 La 'arqueología' de la sustentabilidad en la concepción del patrimonio cultural *Revista de Estudios Sociales* 75–01 de janeiro de 2021, pp 71–86 https://journals.openedition.org/revestudsoc/49281#tocfrom2n6

Castriota L B 2009 *Patrimônio Cultural—Conceitos, Políticas, Instrumentos* (São Paulo: Annablume. Belo Horizonte, IEDS)

Ferrão B 1999 O Conceito de Património Arquitectónico e Urbano na Cultura Ambiental Vimaranense *Candidatura do Centro Histórico de Guimarães a Património Mundial* (Câmara Municipal de Guimarães) https://www.cm-guimaraes.pt/cmguimaraes/uploads/writer_file/document/806/470413.pdf

Foster N 2003 *Architecture and sustainability* https://fosterandpartners.com/media/546486/essay13.pdf

Meadows D H, Meadows D L, Randers J and Behrens W W III 1972 *The Limits to Growth* 5th edn (New York: Universe Books)

Moreira R J 2004 Cultura, sustentabilidade e poderes assimétricos: uma narrativa sobre a renda da natureza na contemporaneidade. *ANPOCS 2004 (Caxambu, 26–30 Oct 2004)* (São Paulo: ANPOCS - Associação Nacional de Pós-Graduação e Pesquisa em Ciências Sociais) https://www.anpocs.com/index.php/encontros/papers/28-encontro-anual-da-anpocs/st-5/st18-4/4043-rmoreira-cultura/file

National Trust for Historic Preservation 2011 *The Greenest Building: Quantifying the Environmental Value of Building Reuse* https://forum.savingplaces.org/viewdocument/the-greenest-building-quantifying

Neto M J B 2001 *Memória, Propaganda e Poder—o restauro dos monumentos nacionais (1929–1960)* (Porto: FAUP Publicações)

Pinto T P 1999 *Metodologia para a gestão diferenciada de resíduos sólidos da construção urbana* Escola Politécnica da Universidade de São Paulo, Departamento de Engenharia de Construção Civil. São Paulo Doctoral thesis. http://www.casoi.com.br/hjr/pdfs/GestResiduosSolidos.pdf

Poulot D 2008 Um Ecossistema do Patrimônio ed C S R de Carvalho *et al Um olhar contemporâneo sobre a preservação do patrimônio cultural material* (Rio de Janeiro: Museu Histórico Nacional) pp 26–43

Sachs I 2007 *Rumo à ecossocioeconomia: teoria e prática do desenvolvimento* (São Paulo: Cortez)

Souza C F and Muller D M 2007 *Porto Alegre e sua formação urbana* (Porto Alegre: Editora da Universidade—UFRGS)

Zanirato S H 2016 Patrimônio cultural e sustentabilidade: uma associação plausível? *Revista Confluências Culturais* **5** 200–11

**IOP** Publishing

Sustainable Urban Development
Topics, trends and solutions
**Luís Bragança, Cristina Engel de Alvarez and Luisa F Cabeza**

# Chapter 9

## The role of a detailed microclimate monitoring in developing a cultural heritage resilience plan

**Ilaria Pigliautile and Anna Laura Pisello**

Cultural heritage (CH) sites play an important role not only in preserving cultural identity, but also in supporting the sustainable development goals of a community and contributing to its economic growth. The effects of the ongoing climate change threaten CH conservation, and even its promotion, if we specifically focus on outdoor sites. The definition of tailored solutions that enhance CH adaptive capability and resilience is of primary importance. The approach to resilience planning in urban areas, in particular for CH, has to be interdisciplinary, since different competencies have to be involved that embrace both humanistic and scientific knowledge. Among others, coupling microclimate monitoring and modelling at different scales of analysis is an important tool for addressing the priority of interventions for CH resilience. This chapter presents advances in microclimate monitoring sensing technologies, with the final aim of suggesting the adoption of comprehensive protocols with which to investigate CH vulnerability to climate change and thus propose effective interventions.

### 9.1 Introduction

The definition of CH encompasses the tangible and intangible attributes of a group or a society that are inherited from past generations, denote our identity, and have to be preserved for future generations (UNESCO WHC 2012). CH sites, such as buildings, historic settlements, monuments, and archaeological sites are physical symbols of our identity that need to be safeguarded from both natural and anthropic threats such as subsidence, flooding, pollution, weathering, urban expansion, etc. The ongoing climate change process is further threatening CH conservation, since it leads to the exacerbation of natural hazards that are going to become more frequent and intense in the near future. The increase in the intensity and severity of extreme weather events draws attention to planning for disaster-resilient cities. Resilient planning concepts could be adapted to CH preservation needs, since the purpose of conservation is to slow down deterioration mechanisms as far as possible, despite the occurrence of external events or changes in boundary conditions.

If we specifically focus on the effects of microclimate on CH preservation, the exposure of artefacts to certain conditions could result in fast material degradation; therefore, proper monitoring of the CH environment, such as its forecast, is of primary importance to prevent irreparable damage. Microclimate monitoring is a fundamental tool for assessing preservation risks and supporting effective counter-actions. While there are already many possibilities for careful monitoring and the control of microclimatic conditions indoors, the opportunities for outdoor action with fine-grained monitoring and tailored interventions are limited. The development of a comprehensive protocol for monitoring the CH status with respect to local environmental variables is necessary as a useful tool for planning effective preventive conservation actions. Such a protocol should include different levels of investigation and analysis, from the long-term to short-term observations made in passing using *in situ* quick assessment techniques. This chapter aims to propose a structure for a comprehensive protocol, starting from the presentation of existing microclimate monitoring methods and of novel sensing technologies that could be integrated to get more detailed information about CH status.

In section 9.2, we present the importance of CH preservation for our communities and how this is going to be compromised by the ongoing climate change, concluding with the concept of resilience for CH. Section 9.3 describes the detail of the role of monitoring, and more specifically of microclimate monitoring, accounting for different spatial and temporal scales of analysis. Finally, section 9.4 focuses on the proposition for a comprehensive protocol for CH monitoring and resilience planning for climate change effects.

## 9.2 The importance of preserving cultural heritage sites

'Heritage is our legacy from the past, what we live with today, and what we pass on to future generations. Our cultural and natural heritage are both irreplaceable sources of life and inspiration' (UNESCO WHC 2012). According to the UNESCO definition, CH are both tangible (man-made or natural) and intangible (e.g. knowledge, performing arts, social practices, oral traditions) resources, inherited from the past and created in the course of time by people and their interaction with the environment. To preserve CH sites means preserving the local identity, unique history, and heritage of a society. Through CH, we can foster our sense of belonging, which further supports community resilience against environmental and socio-economic stresses. This is the case for the Gullah Geechee minority group in the United States, as reported in (Ghahramani *et al* 2020). Indeed, Ghahramani *et al* verified through this extensive review of this minority community that, given the vulnerability of their cultural heritage to climatic and societal changes, their cultural heritage is also the only way to enhance their community resilience and needs to be preserved to promote more sustainable community-led heritage and tourism developments. Therefore, the preservation of CH is additionally a source of economic growth for local communities, since it has the potential to lead the tourism sector and recreation industries by revitalising specific urban areas. From this perspective, Alexandrakis *et al* (2019) proposed a novel socio-economic impact model for analyzing the direct and indirect revenues associated with tourism numbers at CH sites. Through their analysis, they showed that CH provides a range of both market and non-market benefits to society, highlighting the related opportunities for policy interventions for the conservation and promotion of CH, which also need to account for their protection from the impacts of environmental risks.

In the main framework of sustainable development, the role of CH preservation was highlighted by UN in Target 11.4 of the SDGs (United Nations 2020) devoted to making cities and human settlements inclusive, safe, resilient, and sustainable by strengthening efforts to protect and safeguard the world's cultural and natural heritage. Therefore, the preservation of CH is recognized as a tool for fostering the sustainable development of communities. The abovementioned SDG highlights once more the complex nature–culture entanglement (already pointed out by the UNESCO heritage definition) and shed lights on how to reconnect the man-made and the natural environments, in line with sustainability, by explicitly recalling the concept of resilience described in the comprehensive review by Croese *et al*, which addressed 100 resilient cities (Croese *et al* 2020). Resilience is the attitude of adapting while facing adversity and the capability to absorb traumas. Nowadays, one of the main threats to CH preservation is the ongoing process of climate change; thus, CH sites have to become resilient to climate change effects. The following subsections aim to clearly present the main effects of climate (and microclimate) changes on CH and how the concept of resilience has to be implemented for effective CH preservation planning.

### 9.2.1 Climate change effects on CH—the role of microclimate

According to the latest published assessment Report (AR5) from the Intergovernmental Panel on Climate Change (IPCC), there is clear and unequivocal scientific evidence for the warming of the climate system (IPCC 2014). Climate change is already affecting our life from several perspectives, compromising natural, human, and cultural ecosystems. Desertification, rising ocean levels, and the occurrence of extreme weather events (like heat waves, among others) impact the social and cultural aspects of different communities worldwide, affecting their economies, the way they live and work, and how they socialise in buildings and urban areas. Focusing on built areas (historic centres and archaeological sites), the most threatening climate change effects concern the more frequent and more intense occurrences of extreme weather events and the variation in rain patterns that (i) compromise the physical conservation of artefacts and (ii) limit their accessibility to people (both locals and tourists). In order to effectively counteract these two main negative consequences of climate change on CH, it is fundamental to intervene by taking a twofold line of action: (i) to prevent damage to the CH, and (ii) to adapt the CH sites to the ongoing process of change through specifically designed measures that improve the resilience of the area.

Each physical component of a built site is in a dynamic equilibrium with the surrounding environmental forces. A continuous feedback loop operates in between the microclimate and the built environment; they coexist in a delicate mutual-affection relationship. This process is responsible for material ageing and other deterioration mechanisms that cannot be avoided, but they can be slowed down as far as possible through the implementation of proper conservation strategies. The ongoing climate change is altering local microclimates, compromising the dynamic equilibrium between the microclimate and the built environment, and accelerating the CH physical deterioration process. CH conservation is particularly threatened by three main classes of climate change effect: (i) temperature-related (e.g., the increase in the mean annual temperature, superficial temperature increases, and the exacerbation of summer peaks—as recognized at the European level), (ii) water-related (e.g., changes in rain patterns, prolonged heavy rainfall, and flooding), and (iii) wind-related (e.g., changes in the wind field, gusts, and driven particulate matters) effects. In addition to the microclimate impacts, CH conservation is compromised by the exposure of artefacts to specific air pollutants (such as $SO_2$, $NO_x$, and $O_3$) that cause a superficial soiling process by chemically reacting with artefacts' surfaces, and whose increase in atmospheric concentration is attributable to anthropogenic activities.

From this perspective, the main conservation action for those CH artefacts that are preserved in indoors concerns the identification and then the application of the most suitable microclimate conditions through the proper conditioning of indoor spaces. This kind of action cannot be pursued when dealing with built heritage sites directly facing external weather conditions. Here, the only chances for intervention concern the direct sheltering of the most fragile elements, thus protected from direct sunrays or heavy rain exposure, or intervention in the surroundings in order to mitigate negative microclimatic effects while safeguarding the aesthetic of the CH. Possible interventions in the surroundings could involve the limitation of specific anthropogenic activities to reduce pollutant levels in the proximity of the artefacts (e.g. vehicular traffic limitations in historic centres), the introduction of greenery, or innovative superficial finishes that modify the energy balance of the area, reducing the well-known phenomenon of the urban heat island which is characteristic of the microclimate of built areas. These interventions, devoted to microclimate mitigation in built areas, not only have the benefit of limiting the environmental stress on built heritage components, but also of reducing the human-related risks in the area (the health hazard due to overheating can be added on top of those). Thanks to the careful planning of microclimate mitigation actions, both the preservation of physical components of the CH and the accessibility of the area (including its attractiveness for tourists and thus its role in the economic growth of the community) can be enhanced. Therefore, all these actions have the overall potential to enhance CH resilience; this concept is further presented in the following section.

### 9.2.2 CH resilience

Resilience is the ability to withstand adversity, to adapt while facing specific threats or significant stressors. Recent dissertations on the concept of resilience are moving the focus from the notion of bouncing back to a previous state towards the idea of bouncing forward (Holtorf 2018). This changing paradigm involves the concepts of adaptation, learning from past events, and flexibility. This concept has been used in many different fields from psychology to engineering, and in recent decades it has also been applied to complex systems, such as cities. A resilient city is defined as 'a city that has developed the systems and capacities to be able to absorb future shocks and stresses over time to still maintain essentially the same functions, structure, systems, and identity, while at the same time working to mitigate the present causes of future shocks and stresses' (Cloete 2012). Moving from the city scale to the building scale, a resilient architecture can adapt to the changing external conditions by putting in place two fundamental principles: flexibility and adaptability. From this point of view, the built heritage could be an inspiration for contemporary architecture, since it already faced multiple changes to its surroundings and was built according to contextual characteristics and resources.

CH resilience planning has to be driven by risk maps, which highlight the most critical areas and intervention priorities. Two broad main categories should be simultaneously analyzed to assess CH risks: the individual vulnerability of the artefact and the contextual threats. Both of these categories encompass three domains: the environmental, the static-structural, and the anthropic domains. Concerning the artefact's vulnerability, these three domains have to be accounted for by specifically referring to the object of the analysis, to its surfaces and structural characteristics, and to its safety with respect to the artefact's usage. The focus of the analysis is wider when we are looking at contextual danger potentials; the three abovementioned domains express the importance of monitoring and analyzing the CH site's microclimatic conditions and pollutant distribution, geomorphologic characteristics, and demographic and socio-economic descriptors at the regional level. A broad schematic of the risk assessment is summarized in figure 9.1.

Different monitoring and analysis techniques exist for each domain and level of investigation, which must be specifically selected, taking into account the specific requirements of each case study. Furthermore, it is worth noting that all these aspects are interrelated and mutually affected. For example, microclimate conditions impact the structural response of a building (Cavalagli *et al* 2019), but also its accessibility, by threatening visitors' comfort and health (Pioppi *et al* 2019), as better pointed out in the following section.

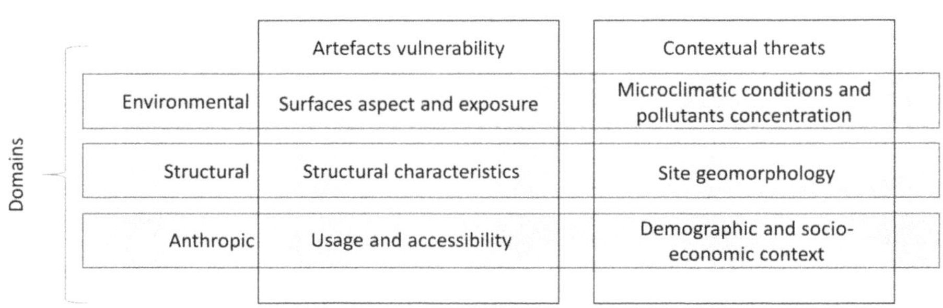


**Figure 9.1.** Cultural heritage risk-assessment schema.

## 9.3 The role of monitoring

The achievement of CH resilience through the proper design and implementation of effective strategies for enhancing the adaptability of heritage sites requires a preliminary action of monitoring the environmental boundaries. The monitoring phase is a required step for the elaboration of risk maps and to set a list of intervention priorities, as better specified in the previous section. A wide range of sensors and techniques are available nowadays, and it is fundamental to define the proper strategy to be implemented in different contexts according to the available resources. Different monitoring techniques can provide different information, presenting different spatial and temporal resolutions. A broad distinction pertains to remote sensing (data typically retrieved by satellites) and in-field monitoring. In-field monitoring techniques can be further distinguished according to the temporal and spatial scale of the analysis: fixed monitoring weather stations can provide long-term data series, but are generally located far from the specific object of the analysis, while portable, miniaturized, inspection devices can get better spatial detail by carefully collecting information in close proximity to artefacts but just during a short-term campaign.

Considering the key role of environmental monitoring and its ability to detect hazards relating to environmental forces, it can be also coupled to other monitoring domains, such as structural damage analysis. The variables that are measurable by means of environmental monitoring may therefore also be applied to detect or co-detect structural damage that should be prevented in order to optimize structural and infrastructural resilience and safety. A possible combined holistic framework that could be put in place, coupling environmental and structural performance assessment and monitoring techniques, has been proposed by Cavalagli *et al* (2019) (figure 9.2).

The following subsections give a more detailed presentation of the in-field techniques and outcome analysis.

### 9.3.1 Long-term in-field environmental monitoring

Fixed weather stations are the proper tools for obtaining long-term in-field environmental monitoring, whose role in CH preservation is twofold: (i) to provide information on the external forces that stress the CH structure and surface materials through the years, and (ii) to support other data sources (e.g., satellite data) and the development of reliable microclimate models. More specifically, the availability of long-term environmental data series would allow us to know the mean extent of the parametric fluctuations that the artefact is generally exposed to, or the occurrences of extreme weather events in the past, with their frequency. This information can be related to the observed CH conservation status and, whenever possible, to its surface weathering and material degradation rate. Additionally, the availability of site-specific environmental data time series can provide better tuning of the microclimate modelling of the area. Microclimate models are useful tools, both for simulating the spatial distribution of the main environmental forces in the CH site and for

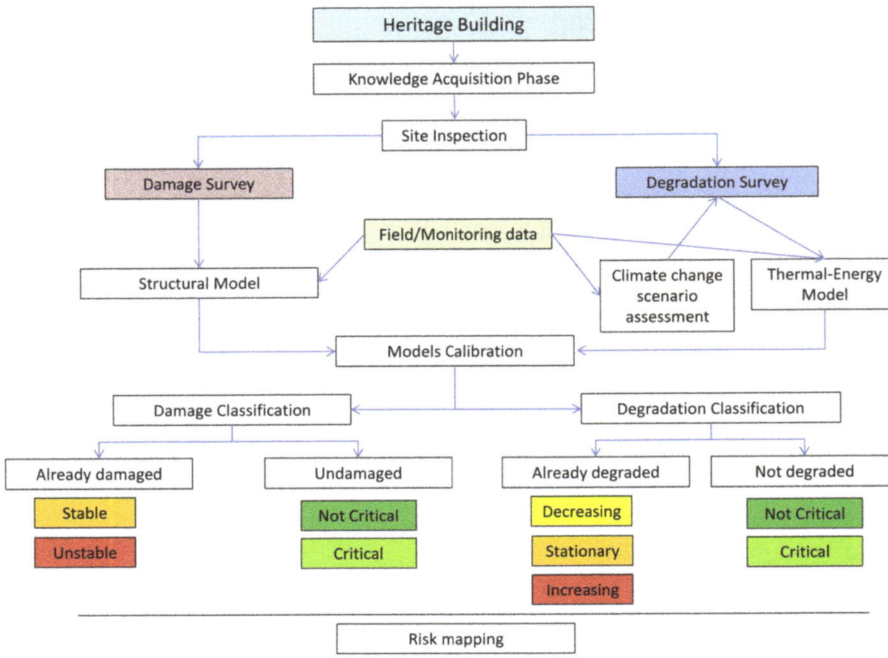


**Figure 9.2.** Cultural heritage operational framework for the detection of microclimate-related risks and resilience indicators which combines thermal energy and structural monitoring and modelling techniques. Reprinted with permission from Cavalagli *et al* (2019).

foreseeing future scenarios according to the climate projections or to microclimate mitigation plans (by checking the effectiveness of planned strategies).

Fixed weather stations integrate various measuring devices combined with data acquisition and processing units. These could be part of a wider network (as for the ones that provide meteorological services) and can provide data in real time or offline. The latter record data on site, internally or through external data storage devices, and the physical intervention of an observer is required to collect the stored data. The environmental parameters most commonly available in fixed weather stations are the following: atmospheric pressure, air temperature and humidity, wind speed and direction, precipitation rate, and global solar radiation. Additional information can be included, according to the purpose of the monitoring. For CH preservation purposes, a set of sensors devoted to air pollution detection is suggested, since this is mainly responsible for superficial CH soiling. More specifically, continuous monitoring of the air particulate matter content, $SO_2$, $NO_x$, and $O_3$ atmospheric concentrations, is suggested to identify whether these values exceed critical thresholds, or if a trend can be recognized that is attributable to specific natural or anthropogenic activities in order to put the proper counteractive strategies in place.

Therefore, a continuously operating complete monitoring setup, has many advantages for the assessment of CH environmental boundaries, the associated danger potential, and the artefact risk according to its vulnerability. Nevertheless, it is difficult to achieve the desired level of spatial detail through fixed weather stations.

The aesthetic preservation artefact itself could limit the possibilities for installing the required set of sensors and data-logging systems where these are really needed (just upon or in close proximity to the object of study). These limitations of long-term environmental monitoring through fixed weather stations highlight the importance of combining this technique with others that are complementary in terms of temporal and spatial resolution. The following subsection presents novel portable miniaturized environmental monitoring systems that can be used for short-term campaigns that focus on the crucial points of the monitored area.

### 9.3.2 Spatially detailed short-term environmental monitoring

CH preservation, the material degradation of artefacts, the surface soiling processes, and also the structural stability of buildings are strictly related to site-specific microclimate conditions and air pollutant concentrations. Fixed weather stations can provide the required information and track the variation of environmental parameters over time, but this equipment is generally located in areas that may not represent the specific conditions experienced by the CH. This is particularly true when taking into account CH complexes, buildings, or monuments that are located within cities, where microclimate conditions below the urban canopy are quite heterogeneous. Miniaturized and portable weather stations are needed to observe intra-urban microclimate variability. Many prototypes have been designed and tested throughout recent decades by the scientific community. These monitoring systems can differ in terms of the number of environmental parameters collected and the type of mobile vector adopted to deploy the tool in the city. The amount and the type of collected data depend on the purpose of the investigation and can range from the simplest configuration for air temperature data collection only up to the most sophisticated systems also including wind field mapping, pollutant concentrations, and directional radiation and/or illuminance data. The hardware complexity not only affects the details of intra-urban microclimate characterization but also the type of mobile vector adopted for these campaigns. Indeed, the most sophisticated setup needs the support of motorized vehicles, such as cars or vans; for example, in the system presented in Kousis *et al* (2021), five monitoring units are involved, each one specifically oriented to directionally record the impact of the urban surroundings. On the other hand, simpler configurations can also be based on bicycles, drones, or directly worn by the observer, as in the experimental campaign carried out in the historical centre of Gubbio, where microclimate heterogeneity was addressed in terms of both its effect on CH preservation and tourist flow enhancement (Pioppi *et al* 2019). Some examples of mobile monitoring systems are depicted in figure 9.3.

The portable monitoring systems that are the most suitable for CH boundary monitoring are the smallest ones, characterized by a high level of compactness and low weight. Indeed, this type of monitoring equipment can be easily brought into CH sites where specific monitoring routes can be designed according to case-specific needs. Furthermore, the possibility of installing the monitoring apparatus on drones offers the opportunity to perform an aerial inspection of the area and to collect vertical profiles of the environmental parameters of interest.

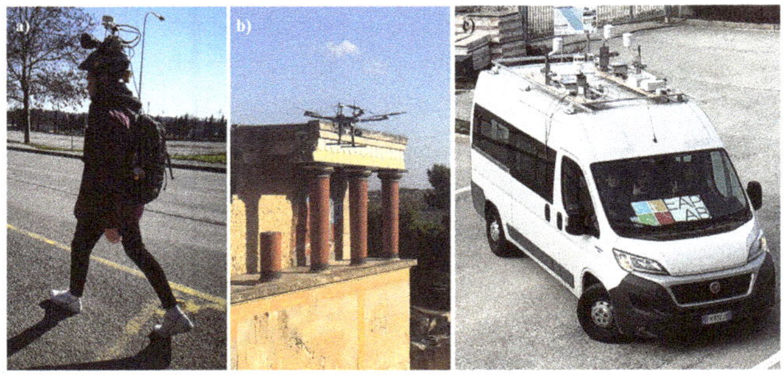

**Figure 9.3.** Examples of mobile monitoring systems (a) directly worn by the observer, (b) installed on a drone, and (c) installed on a motorized vehicle.

In view of these facts, fine-grained spatial monitoring through portable miniaturized systems is an investigation methodology that can be coupled to other more common in-field inspection techniques and is becoming a component of a quick assessment approach which is widely adopted in structural monitoring. More specifically, the quick assessment is a fast investigation approach which is useful for pointing out the critical factors of the investigated object, and for prioritising actions in compliance with observed real needs. By definition, this fast analysis requires *in situ* technologies which have to be simple in use and flexible in their applicability, which are thus adaptable to case-specific needs. Miniaturized and portable weather stations fully address these requirements.

As a further use of this innovative portable equipment, miniaturized microclimate stations allow the analysis of visitor comfort and health risk exposure, mainly in terms of outdoor thermal comfort and air quality. The assessment of the human-related risk component is valuable, since it provides insights not only for CH preservation but also for its promotion among locals and tourists. For example, the observation of critical conditions could suggest the adoption of mitigation strategies aimed at limiting overheating risks (Pioppi *et al* 2019).

A limitation of this approach is related to the short duration of the associated monitoring campaigns. Indeed, environmental monitoring through miniaturized portable systems presents length constraints due to operator availability and CH accessibility. Moreover, the outcomes of this type of monitoring do not only affect environmental parameter mapping along defined routes, but the collected data are both space and time dependent. Therefore, overlong campaigns would require a subsequent data-processing phase devoted to elapsed-time correction in order to provide an instantaneous picture of the observed environmental conditions. To overcome this limit, it is important to combine the spatially detailed short-term monitoring with long-term monitoring from fixed weather stations and satellite data. Additionally, careful planning of the short-term schedule would also provide useful information with which to validate the microclimate modelling of the area; this would allow the inspection of microclimate spatial variability, even under weather boundaries that are not directly related to the short-term campaigns conducted. Finally, it is worth noting that great advances have taken place in electronics and communications technologies in recent years. This trend suggests a near future with a huge amount of publicly available environmental data that could be provided by different data sources for near real-time spatially detailed environmental monitoring, even in compact urban contexts. This perspective further enhances the opportunity to create reliable and effective resilience plans for CH sites based on specific and local data.

## 9.4 Comprehensive protocols for CH resilience

The proper monitoring of CH environmental boundaries is of primary importance in order to develop effective strategies devoted to the enhancement of CH site resilience in accordance with the main purpose of CH preservation. An extensive analysis of CH environmental boundaries would require long-term data series at a fine-grained spatial resolution. As observed in the previous section, different techniques for environmental monitoring produce outcomes characterized by different spatial and temporal resolutions. Nowadays, the two aspects of monitoring duration and spatial accuracy cannot be both maximized through the adoption of one single monitoring tool; therefore, it is of fundamental importance to combine different complementary techniques by putting in place a comprehensive monitoring protocol. Long-term data series collected through fixed weather stations cannot be representative of the real local boundaries affecting the CH, but are useful references for the short-term campaigns involving portable equipment. The careful design and scheduling of the short-term campaigns, to be conducted in close proximity of the artefact, allows environmental information to be obtained at crucial points on significant times of the day and days of the year. The same monitoring path needs to be repeated throughout a year to get information on seasonal and daily trends and extremes. Therefore, the support of continuously monitored data from the closest fixed weather stations is fundamental in order to compare the data provided by these two techniques, to point out the main correlations between the regional and site-specific observed parameters, and ideally to arrive at a continuous picture of local boundaries.

The analysis of the environmental boundaries affecting the artefact, which is complementary to the environmental monitoring (comprising both long-term and spatially detailed campaigns), also has to be addressed through microclimate modelling (Pioppi *et al* 2019). Indeed, microclimate models provide an insight into the spatial distribution of the main parameters and pollutant concentrations, not only under the current conditions, but also with respect to future or alternative scenarios. This investigative tool depends on the availability of monitored data to increase its prediction accuracy throughout a process of calibration and validation. More specifically, the data provided by fixed weather station are more suitable to be inputs to the model, while site-specific data collected using portable stations are more useful for the validation process. Once validated, the microclimate model allows policymakers to assess the impacts of specific interventions for CH preservation in terms of environmental risk and to identify future environmental threats due to climate change.

## 9.5 Conclusions

Moving from the importance of CH preservation and the concept of CH resilience, this chapter focused attention on the existing in-field environmental monitoring techniques used for comprehensive assessment of the environmental boundaries threatening CH conservation. In detail, two main approaches were distinguished: long-term and spatially detailed types of monitoring. Long-term environmental data collection is generally guaranteed by fixed weather stations installed within the region where the artefact object of the study is located. Temporal data series provided through long-term monitoring allow the comprehension of environmental stressors that occurred in the past, together with the main yearly, seasonal, and daily trends experienced. Therefore, these data could be associated with the current conservation status of the CH. On the other hand, a careful on-site inspection of the environmental boundaries through portable monitoring systems measures the specific current conditions and a higher level of detail of the environmental threats faced by both the artefact and people visiting the site. Furthermore, recent trends in sensing and communication technologies suggest that there is an opportunity to relay a huge variety of data that are potentially useful for on-site environmental inspections and for improving the spatially detailed analysis.

The potential and limitations of both fixed and portable environmental monitoring stations have been pointed out, and a comprehensive protocol was finally presented that combined the two techniques and microclimate modelling as a further tool of analysis for predicting the impact of the ongoing process of climate change and the effects of planned mitigation strategies. Indeed, the overall process aims to direct the design of genuinely needed interventions in the CH and its surroundings in order to enhance its resilience and its ability to adapt to surrounding stimuli, such as environmental ones. Finally, it is worth remembering once more that the environmental monitoring is only a component of the CH inspection needed to comprehensively safeguard its preservation. Other kinds of inspection are required, such as the continuous monitoring of CH structural conditions, or the chemical analysis of specific CH samples in the laboratory with the final aim of producing a comprehensive assessment of vulnerability and risks associated with a specific CH, accounting for all the domains of investigation. All these different aspects have to be kept in mind to monitor and promptly act on site to preserve CH and consequently to also improve people's quality of life, stimulate the local economy, and enhance the local environment.

## Acknowledgement

Ilaria Pigliautile acknowledges the H2020 GEOFIT project, G.A.792210 (Deployment of novel GEO thermal systems, technologies and tools for energy efficient building retrofitting).

## Bibliography

Alexandrakis G, Manasakis C and Kampanis N A 2019 Economic and societal impacts on cultural heritage sites, resulting from natural effects and climate change *Heritage* **2** 279–305

Cavalagli N *et al* 2019 Hierarchical environmental risk mapping of material degradation in historic masonry buildings: an integrated approach considering climate change and structural damage *Constr. Build. Mater.* **215** 998–1014

Cloete C E 2012 Assessing urban resilience *WIT Trans. on Information and Communication Technologies.* 44 (Southampton: WIT Press) pp 341–51

Croese S, Green C and Morgan G 2020 Localizing the sustainable development goals through the lens of urban resilience: lessons and learnings from 100 resilient cities and Cape Town *Sustainability* **12** 550

Ghahramani L, McArdle K and Fatorić S 2020 Minority community resilience and cultural heritage preservation: a case study of the gullah geechee community *Sustainability* **12** 2266

Holtorf C 2018 Embracing change: how cultural resilience is increased through cultural heritage *World Archaeol.* **50** 639–50

IPCC 2014 *Climate Change 2014: Synthesis Report. Contribution of Working Groups I, II and III to the Fifth Assessment Report of the Intergovernmental Panel on Climate Change* (Geneva: IPCC) https://ipcc.ch/site/assets/uploads/2018/02/SYR_AR5_FINAL_full.pdf

Kousis I, Pigliautile I and Pisello A L 2021 Intra-urban microclimate investigation in urban heat island through a novel mobile monitoring system *Sci. Rep.* **11** 9732

Pioppi B *et al* 2020 Cultural heritage microclimate change: human-centric approach to experimentally investigate intra-urban overheating and numerically assess foreseen future scenarios impact *Sci. Total Environ.* **703** 134448

UNESCO WHC 2012 *UNESCO World Heritage Centre: Tajikistan*

United Nations 2020 The Sustainable Development Goals Report 2020

# Chapter 10

# Designing buildings for adaptability, flexibility, and durability

**Rand Askar, Luís Bragança and Helena Gervásio**

Our growing cities persistently face emerging demands for housing, work, health, and mobility, among others. Yet buildings are often designed and set up to embrace one function, such as an office building, a flat complex, or a shopping mall. What happens when users' preferences change and new societal needs emerge? These mono-functional buildings are transformed into outdated or even obsolete structures, bringing about high building vacancy rates and consequently premature demolition. The demands of global sustainability induce us to shift our design culture towards embracing a new vision, in which buildings are created to be flexibly changed in response to the ever-changing societal needs and preferences. Designing buildings with an enlarged adaptive capacity for the variables that will arise over their lifecycle is therefore becoming a primary necessity, as we know that change is inevitable. The inclusion of adaptability in buildings is acknowledged to be an important design criterion that addresses potential built-environment challenges in a sustainable manner. Adaptability is usually referred to as the capacity to be modified to fit new conditions. This indication sounds simple and explicit enough to be understood. Nevertheless, literature studies have revealed diverse interpretations and reflections on links with other concepts (e.g. flexibility, durability, open building, and building layers) and contemporary trends, such as the role of the circular economy in building and climate change resilience. This chapter discusses the key principles and features of the adaptability of buildings and building components, exploring the different interpretations and identifying efficient principles and strategies in the light of promoting concepts and underlying models. The chapter concludes with a confirmation of the inevitable need to incorporate adaptability into the criteria when designing buildings.

## 10.1 Why adaptable buildings?

Over the last century, Europe's building stock has grown tremendously. Global construction markets have rebounded after years of poor growth, resulting in a

doi:10.1088/978-0-7503-3971-1ch10

strong upturn in the world's building stock, which represents the largest financial, physical, and cultural asset in the industrialized world. Our growing cities persistently face emerging demands for housing, work, health, and mobility, among others. Meanwhile, in an unprecedented way, they are coping with challenges and accelerated trends including increased urbanization, technological innovations, and climate change. Yet buildings are often designed and set up to embrace a defined function and a fixed form, such as an office building, a flat complex, or a shopping mall. This is the reason that they end up as obsolete objects when user needs change and new societal demands arise. Building obsolescence resulting in high vacancy rates is the leading motivation for major refurbishments and premature demolitions, which both incur elevated costs and additional energy input and generate massive amounts of waste that is only partially reused or downcycled into lower-quality products.

Although the building industry is increasingly becoming a key pillar of the business cycle and the economy, it is acknowledged worldwide to be a significant contributor to ecological degradation. The construction sector alone accounts for around 33% of greenhouse gas emissions, 40% of material utilization, and 40% of waste production (Ness and Xing 2017). The intensified pressure on resources which are becoming progressively scarce has largely contributed to inflating overall costs and has created uncertainty in the short term. The relation between environmental problems and climate change on the one hand and building-related activities and practices on the other has brought issues of resource valorization and environmental impact minimization to the top of governments' and scientific societies' agendas.

In response to the above arguments and figures, many studies have approved the inclusion of adaptability in buildings as an important design criterion that addresses potential built-environment challenges in a sustainable manner. The realization of adaptability in buildings has always been of constant interest for scholars and practitioners across multiple disciplines including architecture, engineering, construction technology planning, and management, where adaptable buildings are largely perceived as intrinsic to sustainable construction. This stems from their potential to avoid premature demolition and the costs associated with investing in new materials while preserving the value of materials and embodied energy. This also implies an extension of the building's useful life, since alterations can happen while preserving a significant part of a building. Resource scarcity and environmental degradation caused by building-related activities, together with the huge investment in refurbishment projects in EU countries, increasingly drive the need to achieve adaptability in buildings as a means of reducing their ecological footprint, extending their economic viability, and minimizing their maintenance costs.

The future of buildings is surrounded by uncertainties of multiple types, making change inevitable over their lifecycle. The impossibility of predicting the future further emphasizes the value of adaptable building design in addressing factors of change which can be related to social and local aspects (e.g. user preferences, cultural demands, existing materials), environmental motives (e.g. natural hazards, climatic changes, such as heat waves), technical requirements, and functional performance (e.g. to embrace technological improvements), economic factors, legislative issues (e.g. regulations and policies), stakeholders' interests, etc. Static permanent buildings are therefore no longer an option in our modern and future life. Conversely, a realizing sustainable future depends more than ever on the capacity of those buildings to adapt to ever-changing operational conditions and preserve their functions to meet the emerging needs of users.

## 10.2 Featuring adaptability in buildings

The notion of adaptability in the built environment is quite vague and is often confused with other terminologies such as: flexibility, polyvalence, and versatility, among others. The widespread usage of the word across multiple disciplines and the different settings where it applies make it rather difficult for built-environment scholars to agree upon an exact definition of the concept. Yet the contextual conditions usually help in determining the type of adaptation that applies when asking 'adaptable to what?' The answer often lies in the type of change desired or the action needed to address a change that occurred under previous operational conditions.

In general, adaptability can be defined as 'the capacity of a building to change in response to the emerging needs or varying contextual conditions, prolonging therefore its useful life while preserving the value for its users'. This definition, like other adaptability definitions in the literature, assumes that a building has the ability to evolve over time by manifesting changes stimulated by external or internal variables (Askar *et al* 2021). The ability to accommodate change—regardless of the essence of that change—is an overall feature of adaptable buildings, which can be seen in state-of-the-art adaptability studies. Change of use is the most common type of change discussed by scholars. This is probably because users themselves can frequently perform changes of their space uses with small non-professional interventions in the interior. Still, some changes of use or function can imply larger and more complex interventions. However, other types of change are also identified in the literature, such as: spatial changes, layout reconfigurations, changes in size or volume, changes in function or use, changeable parts or components, and performance changes.

The usage of 'adaptability' as a buzzword in the discourse about building obsolescence and redundancy stems from associating it with the idea of change in its multiple forms, making it an umbrella concept for all notions and strategies that imply a certain type of change. These notions can be therefore perceived as dimensions or types of adaptability. In this regard, the literature reflects on a plethora of strategies that contribute to the physical capacity of a building to change and therefore adapt, such as: extendibility or scalability, flexibility, recyclability, reusability, transformability, upgradability, convertibility, durability, etc.

Assuming the presence of multiple variables over a building's lifetime makes it crucial to transform the usual focus of form and function on the actual needs into the embracement of a context- and time-based view of long-term changes (Schmidt *et al* 2010). The time factor is an essential enabler for change here. Buildings are usually set up as static rigid objects, which neglects the temporal reality of architecture created by the variables of a certain context. Adding the time factor to the equation allows a contemplation of the building as a dynamic system that interacts with emerging demands, resulting in change over time. The dynamic interaction creates the ability to maintain an extended useful life, which cannot happen with a one-time solution but rather with continuous possibilities over time. However, an extended useful life means that there is a persistent value for users. In all perspectives, people are always part of the equation, as buildings change to accommodate their changing needs, provide them with better service, and ensure their comfort and protection. The value generated by efficiently addressing needs is often lost in traditional building, as inevitable mismatches develop over time between the supply of space and the demand for it (Leaman *et al* 1998).

The capacity of a building to respond to emerging changes stems from the early design decisions made in the first place (Schmidt *et al* 2010). These decisions are viewed in terms of their relevance to particular contextual conditions. However, they usually disregard the fact that that the context does not have a stable nature but is rather one of continuous evolution over time. Analyzing contextual variables therefore helps to anticipate the changes and adaptations that might be needed in the future. Nevertheless, a series of uncertainties still might occur, creating scenarios other than those expected from a consideration of the context.

Given all that, delivering a genuinely adaptable building calls for the right mix of time-based, context-oriented perspective with the capacity to adapt to new conditions, thus extending the useful life of the building while creating continuous value for its users.

## 10.3 Connecting adaptability, flexibility, and durability

Despite the simplicity of the 'adaptability' definition, the literature for the built environment shows multiple interpretations and reflects a high level of relevance to other concepts and terminologies. The term 'flexibility' is probably the one most associated with the concept of adaptability. The two concepts keep appearing together in the literature; some studies suggest their use interchangeably, while others attach distinct characteristics to each of them. However, this relies to a large extent on the context in which they are applied. Still, flexibility is often associated with small, quick, and short-term changes, while adaptability refers to considerable large-scale changes over long-term periods (Leaman *et al* 1998). Furthermore, several studies have related flexibility to changes in the configuration of interior spaces which can be performed by the users, while for adaptability, a professional guided intervention, often carried out by an outsider (e.g. the designer), is required. Generally, the perception of flexibility as a concept or strategy contributing to building adaptability is dominant when addressing change. We perceive adaptability as a complex concept that goes beyond merely improvising solutions to address an emergent need; it rather relies on a mix of instant response and a series of inherent properties anticipated by design and facilitated by process variables (stakeholders, regulations, owners, etc). Nevertheless, flexibility can contribute to a certain degree of adaptability, proposing limited alternatives to, or often partially addressing, a certain change.

Adaptability aims to extend the service life of buildings by facilitating changes of their use, size, performance, etc. The reason that many practitioners consider designing durable structures is that it is a major way of achieving adaptability. What is meant by durable structures here is those that have the capacity to support multiple loading scenarios and therefore a variety of functions. Changes in building function often entail changes to the required design loads. Structures should be therefore designed to sustain the worst-case scenario. For this reason, durability and adaptability are usually considered to be complementary concepts that support the mutual objective of extending the lifetime of materials and technology in buildings while preserving usefulness for their users.

The relationship between the three concepts (adaptability, flexibility, durability) is always present in mechanisms and models of application, which will be seen in the following sections.

## 10.4 Models and trends

The overall perception of adaptability starts from the recognition of a building as a dynamic system consisting of a flexible object and a flexible process of design, construction, change, deconstruction, and reconstruction (Askar *et al* 2021). This understanding replaces the conventional conception of a building as a finished, rigid object with a defined function and form.

However, the broad scope of this concept of adaptability in buildings is translated into practice through strategies and ideas introduced by comparable concepts for similar aspirations, such as the 'open building', 'shearing layers', resilient design (a key theme in the Sustainable Development Goals (SDGs)); and circular economy principles. These resonant concepts and their related principles have helped to further shape the concept of 'adaptability' and enable the development of its strategies.

### 10.4.1 The 'open building' movement

The 'open building' movement, which propagated back in the 1960s, is often suggested to be the origin of the concept of adaptable design. The tremendous increase in social housing demand after WWII created a strong catalyst for the spread of such a user-empowering movement which had the aim of delivering more efficient housing units that addressed people's needs.

The concept of open building was promoted by Habraken, who, in his book 'Supports: An alternative to mass housing' (1972) criticized the identical design and architectural character shared by many of the housing units set up in post WWII reconstruction, which ignored social diversity by implementing decisions that residents should have made for themselves, according to their evolving demands. Since 1972, Habraken has emphasized the need for distinguishing between two types of building element: the 'base building', which is the structural support that is subject to investor decisions, and the 'fit-out', which is the flexible infill to be modified by inhabitants over time (Geldermans 2016). The base building consists of the durable part that is the fixed structure, while the infill or fit-out consists of the flexible parts that are exposed to frequent changes. The connection between the base building and the fit-out is achieved in a way that minimizes interface problems, which allows a building to adapt to the users' needs over time. This is achieved by designing functionally independent systems with detachable interfaces.

The ideology of 'open building' has been considered by several studies as the basis of create adaptive design and reuse strategies (Ness and Xing 2017). The technical feasibility of the concept has been examined using few practical examples. The Japanese experience is an example worth mentioning here due to the huge efforts involved, despite not achieving considerable success (Schmidt *et al* 2010). The Japanese implementation of the concept was reflected in the adoption of rectangular plans for their houses. This allowed easy interior subdivision using moveable partition systems of sliding interior panels and foldable partitions (Sinclair *et al* 2012). The Japanese further contributed to the open building idea by creating distinct ownership patterns split between the 'base building' and the 'fit-out'.

The open building idea has been progressed by other studies to extend the flexibility of use to structural elements, which has been found to be fundamental to extending the technical lifespan of buildings. However, the realization of structural flexibility should be examined in such a way that it does not hinder the infill flexibility but rather facilitates it. For instance, using moveable columns according to a certain grid allows more options for layout arrangements. Extending adaptability to engage with the flexibility of structural elements helps in the development of more comprehensive solutions to address change. However, structural flexibility can only be achieved if it is preconfigured from the design phase onward, otherwise, the technical feasibility and cost viability would be in question.

Given all that, the practical application of open building has been limited to residential buildings, mainly located in Japan and the Netherlands. The Dutch experience was oriented towards social housing developments, in which plans were divided into equal-sized multifunctional spaces. The spaces were used as bedrooms at night, living rooms in the evenings and at weekends, and study spaces during the day for family children and young members (Sinclair *et al* 2012).

The idea of an incomplete housing unit that takes the form of a structural frame with free space to be personalized by the user has been advocated by several scholars as well as practitioners, such as the architect Alejandro Aravena, who implemented an incremental housing methodology to create affordable and accessible housing solutions that simultaneously guarantee user satisfaction (Askar *et al* 2019). His project in 'Quinta Monroy' in Chile has been an outstanding and successful example of the adaptability of structure and interior space (presented in figure 10.1). He referred to the process he utilized in the project as 'infrastructure as housing'. His idea was to construct two-story half houses with a space left between houses for future expansion. Each half house delivered the core of a home to be finished by the residents and contained all the basic and essential elements and utilities which cannot be built by the residents, such as the supporting structure, stairs, bathrooms, and kitchens. The rest of the elements: partitions, interior finishes, and the remaining enclosure were left for the inhabitants to provide according to their resources and necessities.

In view of all the above, the concept of open building has mainly focused on the idea of separation between structural and non-structural building components in

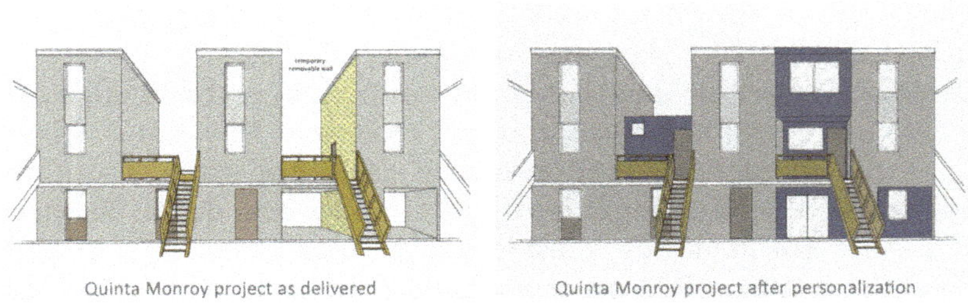


**Figure 10.1.** The incremental housing methodology in the Quinta Monroy project by architect Alejandro Aravena.

order to deliver insights into design and process flexibility. The early endeavors were clearly focused on spatial flexibility and layout reconfiguration for residential use only. However, the concept did not identify clear criteria for the evaluation of the elements of flexibility and overall adaptability. For example, it did not comprehensively address the connectivity and interdependencies between the elements respectively comprising the base structure and the fit-out, which include elements with different service lives.

### 10.4.2 Building layering and decomposition

Buildings are inherently complex objects, since each of their materials and components has a distinct lifecycle, yet all interact dynamically in a certain space and time to perform certain functions. The static integration of elements in closed structural systems neglects the distinct levels of durability of each of a building's components and systems and brings them into rigid incorporation within the fixed space of a building that is expected to have a long life. The extensive lifespan, together with all the changes that occur during the service life of a building, make it rather complicated to anticipate future scenarios. Furthermore, common dependencies between systems prevent any change, as they would entail consequences for the entire building's integrity.

The dependencies existing among system elements and their influence on the mechanism of adaptability have been widely explored in the literature. A number of methodologies were therefore proposed by scholars to explore what composes a building and what kind of interrelationships and interdependencies exist between its systems, components, and layers. Investigating the complexity of a building's lifecycle and analyzing its elements' lifespans make it possible to determine how a building can change and what parts can enable this change.

The model most discussed in the literature is the concept of building layers, which relies on the consideration of a building as a system composed of constructed layers defined by their varying lifetimes and functions. The concept of layers was first discussed by Duffy (1990), who considers the perception of a building as a rigid constant object to be invalid. He states '...there isn't such a thing as a building...a building properly conceived is several layers of longevity of built components' (Duffy 1990). Duffy identified four layers in buildings, according to their rate of change: the shell (structure, 50 years), services (e.g., heating, plumbing, 15 years), scenery (fittings, decorations, 5–7 years), and set (e.g., furniture, daily). Later, Brand (1994) suggested an extended approach of 'shearing layers', which consists of six layers with different lifespans (site, structure, skin, services, space plan, and stuff). Brand's model assumes that the components of the service (e.g. electrical cables) will need to change on an average timescale of 7–15 years. They should therefore be separated from longer-life components, such as structural columns, which have an average service life of 30–300 years, as well as from shorter-life elements, such as partition walls, which belong to the space plan layer (3–30 years). Many proposals were later built upon Brand's shearing layers of change; some added more layers with different lifespans. However, the later proposals approached the mechanisms of change more comprehensively through developing configuration criteria and defining the arrangement and relationships between elements and components to ensure a certain level of independence and exchangeability among them, for example, by classifying components based on their characteristics or their tendencies to change. Further developments in the areas of understanding building composition and facilitating change were made through the categorization of products into platform-based families, which helped to identify the commonality potential within each product family and the feasibility of standardizing systems and subsystems. In this regard, multiple types of categorization were developed using product families based, for instance, on the hierarchy of product architecture or on the system layout. Schmidt *et al* (2011) proposed to address the effects of certain components and subsystems on the overall capacity for the transformation and reconfiguration of a building during its lifespan. This also entailed questions about the way in which building components are clustered and the level of dependency that exists between components of varying lifespans. To address this shortcoming, Schmidt *et al* (2011) proposed to employ the design structure matrix (DSM) method, which incorporates the two concepts of decomposition and dependencies to deliver quantitative results. In order to deliver novel insights into the design process for adaptability, they used the DSM method to group building components into Brand's layers, then understood the interrelations between them, and after that, examined the potential reorganization of the components through clustering.

The concept of categorizing building elements into temporal layers helped to change the general perception of buildings as static objects toward examining them as dynamic systems with internal and external interactions. Investigating the elements that compose a building and identifying the temporal layers to which they belong deliver a key method for incorporating an adaptive capacity into systems. This is achieved by allowing flexibility and changeability in shorter-life layers and components, while ensuring the durability of longer-life layers and components. However, proactive maintenance planning should be included in order to achieve the desired outcomes of adaptability and guarantee straightforward access for the assessment of components of different lifespans to determine their functionality and facilitate their replacement when necessary. The assessment can be performed by identifying specific access points while taking care not to damage the components next to the replaced one. However, a clear identification of the components and materials that compose each of the shearing layers should be further specified (Geldermans 2016).

### 10.4.3 Design for resilience

Our world is unprecedentedly witnessing an acceleration in climate change and environmental degradation. Enhanced global warming is disrupting climate patterns across planet Earth, leaving urban areas and their elements at risk of confronting more natural disasters (e.g. tsunamis, floods) and heat waves, among other critical threats and challenges. As climate change cannot be reverted in the short term, it will persist in being a reality for the next few decades. An efficient response to the changing climate would therefore call for an enlargement of the capacities of our built-environment systems to respond to uncertainties and future disruptions (Askar *et al* 2021). In this regard, fostering so-called 'urban resilience' in the face of these environmental and socioeconomic threats has gained international attention at both the academic and policy levels, being a fundamental goal for cities and built environments (Sustainable Development Goal N.11: make cities and human settlements inclusive, safe, resilient and sustainable).

Although the concept of resilience often refers to future extreme climate events, such as natural disasters, and climate adaptive survivability in their aftermath, a methodical integration of resilience into systems is required to cope with chronic stresses and gradual trends, such as resource shortage and rises in average temperatures, which can lead to greater issues if overlooked. This idea stems from the two main directions of climate resilience research and application: mitigation and adaptation. While mitigation deals with the causes of climate change by seeking radicalization and diversification strategies to minimize GHG emissions (e.g. by minimizing resource and energy consumption), adaptation targets the effects and consequences of climatic disruptions by enhancing the adaptive capacity of systems to absorb their impacts while continuing their key functions (Askar *et al* 2021). Still, the two directions are complementary and intrinsically linked; for instance, achieving long-term adaptation depends—in large part—on incorporating mitigation measures. On the other hand, implementing mitigation strategies alone does not prevent acute events from happening. Given that, realizing long-lasting urban sustainability and fostering resilience to climate change require frameworks that combine both adaptation and mitigation.

Enabling an adaptive capacity in buildings is recognized as a key method for achieving a resilient built environment. Designing buildings for adaptability is aligned with 'design for resilience', as both share the objective of addressing problems related to environmental management in the built environment. This relationship is strongly reflected by the literature, in which the concept of adaptability has always been perceived as an important factor that contributes to both aspects of resilience. The adaptive reuse of buildings is a strategy of recycling resources in place which largely contributes to the current climate change mitigation agenda. Also, the adaptive capacity of buildings to change their performance in response to climate conditions is an essential method for achieving adaptation strategies for resilience.

In the scope of a built environment, 'urban resilience' can be defined as 'the ability of an urban system—and all its constituent socio-ecological and socio-technical networks across temporal and spatial scales—to maintain or rapidly return to desired functions in the face of a disturbance, to adapt to change, and to quickly transform systems that limit current or future adaptive capacity' (Meerow *et al* 2016, p 2). This definition, together with others throughout the literature, acknowledges the importance of change as requisite to reach a resilient status. The need for 'change' is also an overall feature of adaptable building systems which perceive climate as an important change factor to be taken into consideration to address future uncertainties. However, resilience thinking delivers valuable insights into the process of change that manifests an adaptive capacity in the system to address changes in the built environment, particularly predicted ones. Achieving a resilient design is not a single standard solution or perspective. It rather relies on translating the different perspectives of a scene into proactive and reactive measures through mitigation and adaptation, respectively, to deal with internal and external stressors. However, given the temporal and non-static realities of future building contexts, achieving resilience measures in those is a challenging task. Incorporating adaptability and flexibility features is therefore a key factor in creating time-based and context-oriented solutions which significantly contribute to reducing the vulnerability to future anticipated and unexpected impacts.

The coherent relationship between resilient design and adaptable building is further manifested through the characteristics of resilient systems which clearly contribute to, or facilitate, adaptable design strategies and vice versa. Table 10.1 presents the five main characteristics of resilient systems (robustness, rapidity, redundancy, reflectiveness and passive survivability) with their correlation to adaptable building design.

### 10.4.4 The circular economy in buildings

In response to the elevated ecological footprints resulting from industrial activities, the circular economy (CE) has emerged as a novel model of innovative application for enhanced sustainability; it aims to disengage economic prosperity from resource consumption. The CE concepts suggest a systemic shift from the current linear model of economy (take, make, dispose) by adopting a circular model of closed-loop value chains. This concept is becoming more relevant for academia, policymakers, and non-governmental institutions. In 2015, the European Commission released an action plan to promote the implementation of CE in various sectors, including the building sector. The European package for the circular economy considers the building sector to be one of the five priority sectors in which to foster the CE approach by developing new eco-technologies and circular models for all building-related activities and processes (Askar *et al* 2021). Rethinking building operations while considering circular economy principles promises to significantly contribute to resource efficiency and GHG emission reduction, given the considerable contribution of the sector to the negative environmental footprint.

'Design for adaptability' (DfA) is an important strategy promoted by the CE to address emerging design challenges and achieve circularity in buildings. DfA aims to

Table 10.1. Key resilient system characteristics and their relation to adaptable design strategies.

| Resilient system characteristics | Description | Correlations with adaptable design | Application strategies |
|---|---|---|---|
| Robustness | The strength and capacity of the system to sustain and absorb unexpected disruption without major damage or functional failure | Robustness ensures a structural durability which facilitates the switch between multiple uses and loading scenarios | Reinforcement of existing foundations and structure |
| Rapidity | The capacity of the system to rapidly address a certain need or recover from a disrupting event; this is achieved by the flexibility of the system to reconfigure itself and reach a new stabilized status | Adaptability, being a proactive attribute inherent in a system, allows for rapid changes | Clear story height provides rapid solutions for functional changes that require higher ceilings |
| Redundancy | The spare capacity in a system to handle an emergency through achieving a desired function or meeting a certain need in multiple ways under a certain stress or during a severe event | Redundancy creates overcapacity in the system, which largely supports change scenarios, making buildings more adaptable | Backup generators, cavity floors, the use of renewable energies |
| Reflectiveness | The capacity of the system to continually evolve, modifying its original state in response to varying conditions and needs | Adaptability as an inherent attribute of the system does not provide a permanent, one-time solution; it rather reflects on the emerging conditions to deliver a customized answer whenever needed | Adaptive envelopes and building skins maintain interior thermal comfort by reflecting on the external climate and weather conditions |
| Passive survivability | The capacity to operate with minimal external inputs and withstand an extended loss of power while maintaining livable conditions; this is achieved through passive survival devices or measures | Employing passive measures in building design contributes to increased adaptability to climatic changes | Green roofs, double-glazed windows, passive cooling systems, optimal orientation, renewable energy systems |

prolong the useful life of buildings, their components, and materials by employing lifecycle thinking to address issues of building obsolescence. DfA and the adaptive reuse of building components are strongly linked to the CE paradigm, due to their coherent relevance with the main CE principles: 'closing the loop' and 'slowing the loop'. Closing the loop happens through intensified reuse and upcycling, aiming to retain materials and components in a continuous loop of value chain. On the other hand, slowing the loop assumes an extension of the longevity of materials and products, keeping their value, quality, and performance as high as possible for as long as possible. Both principles ensure a minimization of the demand for energy and new material inputs, hence decreasing the environmental impact associated with lifecycle phases, particularly the end-of-life phase.

Since many of the service factors that cause obsolescence are linked to building design (damaged components, poor design/space/planning/construction, space requirements, user preferences and needs, etc.), designers are therefore increasingly encouraged to depict buildings that are flexible enough to incorporate changes with minimal extra resources and waste generation over their service life. Design choices play a central role when targeting circularity in buildings. Design is the way that buildings are created and assembled. It defines their functions, uses, and levels of adaptability or obsolescence. It also determines the ways in which materials and components are used and interconnected to perform as an integrated entity. Designing buildings for adaptability interacts with these system configurations to deliver a dynamic alternative to the static objects delivered by current design patterns. However, the perception of building adaptability is extended by the CE to embrace new ways of owning and managing properties, creating therefore new business model opportunities for different ownership patterns of building products. This is an important contribution to ensure economic growth. For instance, adaptability from the CE perspective is also an efficient approach to the design and delivery of multipurpose facilities for shared and alternated use which seek to retain more value from long-life structures in a full-lifecycle approach by intensifying their utilization and maximizing their service life (Ness and Xing 2017).

In a circular economy, preserving the value of products for as long as possible is considered to be fundamental. This idea is derived from the strategy of slowing the loop, which, from a design point of view, can be implemented through adaptable design that implies an extended service life for a building and its components. However, the realization of long-life building is not viable without developing evidence for the efficiency of the end-of-life options. Therefore, according to the circular economy, design for adaptability goes hand in hand with the other circular building strategies, such as design for disassembly and deconstruction, design for reuse, space transformability and trans-functionality, and design for longevity, among others. For buildings to adapt to evolving demands, they should be designed not only to permit reconfiguration during their life, but also to accommodate the reuse, modification, and replacement and recycling of redundant components (Schmidt *et al* 2011).

The following paragraphs explain this synergic relationship between DfA and three CE-promoted strategies: design for longevity and durability, design for deconstruction/disassembly and standardization and modular design.

- Design for longevity and durability. Longevity means an extended life, while durability refers to a strong structure that has the capacity to cater different load scenarios. Strategies for longevity and durability stem from the principle of slowing the loop, as they entail a reduced demand upon virgin resources and energy. They are therefore applied to foster circularity in the built environment. Meanwhile, they are coherently linked to adaptability. This is because achieving a durable structure with an extended lifespan would be pointless if it did not deliver a useful function. Conversely, combining structural durability and flexibility of use generates higher values of circularity. Longevity and durability can be applied to a whole building design and at the component and material scales. Still, for components and materials, these strategies might be deemed less efficient, as materials often lose quality over time, which restricts the number of times they can be directly reused, eventually creating the need for recycling which engages extra resource and energy input. As for buildings, longevity and durability can be achieved by replacing damaged or inoperative components as needed. Again, this counts on the adaptive capacity of the building system to facilitate the replacement. The adaptive capacity allows a certain level of independence, in which elements of different functions and lifespans (the shearing layers' concept) enable easy separation. Nevertheless, realizing synergies between the strategies of DfA and longevity should necessarily imply a lifecycle consideration in order to retain the maximum value of products and materials at their end-of-life phase.
- Design for deconstruction/disassembly. Design for deconstruction and design for disassembly are used interchangeably in the scope of CE. They are both referred to as DfD. DfD is a strategy for designing components and products that can be easily disassembled and recovered from a building at every level or life stage. This allows the reuse of materials and components to their highest extent and in multiple structures. The strategy of DfD is recognized to be an important enabler for DfA and vice versa. This recognition is derived from the consideration of adaptability as an architectural strategy in which certain components can be changed in response to external factors, for instance, emerging user needs or variables of the surrounding environment (Askar *et al* 2021). The synergies between DfD and DfA are perceived by various studies to add value to circularity and eventually sustainability, since they imply reduced costs and ecological impact. This is because DfD goes beyond the active life of a building, addressing the destination of building materials and components by accounting for the end-of-life scenarios of buildings from the early design stages; this generates a further advantage when combined with DfA. Moreover, DfD is a strategy that depends to a large extent on the level of interdependence between components and their accessibility for safe removal and replacement. DfA is essential here to achieve minimized interactions between system components and to create specified access points that facilitate replacement, repair, and recovery. The utilization of demountable connections is also an important enabler in this

regard, as it ensures that components with different functions and lifespan remain independent from each other.

- Standardization and modular design: Standardization and modularity refer to the uniformity of component sizes and interfaces, respectively. These strategies are promoted by the CE, since they allow for the intensified reuse of products and materials in multiple structures without essential losses. In addition, the implementation of these strategies is perceived as the easiest way towards adaptability, as they enable a high level of interchangeability and flexibility in configuration. They also facilitate the disassembly, direct reuse, and replacement of components. Furthermore, the standardization of materials is an important enabler of the CE principle of 'narrowing the loop', which is less examined in favor of the other two CE principles: closing and slowing the loop. Narrowing the loop refers to using fewer resources per product. By this, we mean that standardized materials allow for more efficient recycling. In the light of CE, the strategy of standardization also includes the connections that are the key facilitators of change, ensuring the easy removal and replacement of components. Standardizing connections is of great value, as it exempts components from being standardized (Askar *et al* 2021).

## 10.5 Discussion and final remarks

This chapter highlights the importance of designing buildings for adaptability as a potential response to the issues of obsolescence and redundancy that often result from the temporal reality of buildings' contexts and surrounding environments. This chapter underlined the distinct attributes of building adaptability as an essential criterion of the sustainable built environment due to its major role in minimizing resource and energy consumption. The evolution of the concept and its preliminary perception was illustrated in terms of the open building movement and the later models built upon the principle of separating the structural elements of buildings from the interior infill. However, the practical implementation of these models has been limited to one-off residential projects, where the adaptability of the infill only addressed spatial flexibility and some layout configuration possibilities, but fell short of addressing more complex adaptations and interface issues between systems and elements. This is because the support–infill theory could not deliver a clear evaluation of element flexibility and consequently the overall adaptability. In response to these issues, the concept of shearing layers discussed by Duffy, Brand, and Schmidt (among others) has delivered further insights into adaptability mechanisms, allowing an analysis of building configurations and the interdependences between systems, components, and layers. Layering building elements helps the development of feasible solutions and the identification of fundamental obstacles to the implementation of adaptability by allowing shorter-life layers or components to have flexibility and longer-life layers and components to have durability. The independence of layers also ensured an easy process for the assessment of components with different lifespans to determine their functionality and facilitate their replacement when necessary. Shearing layers, as a theoretical concept, did not deliver practical solutions for intersection zones and the interchangeability of elements; neither did it define a clear characterization of the elements constituting each of the temporal layers. Recently, these issues have been raised by trending strategies prompted by the circular economy principles in buildings, which have delivered advanced solutions reflecting on the concept of layers and adaptability in general. Implementing circularity in buildings sheds a light on the necessity of combining multiple strategies under a comprehensive framework in order to achieve higher levels of adaptability and circularity, thereby contributing to better management of resources. For example, the consideration of design for disassembly principles allows adaptability to act on the structural, spatial, and material levels. The recognizable contribution of adaptability strategies to environmental impact minimization makes them a valuable approach to achieving resilience and dealing with change in proactive and reactive ways.

## Bibliography

Askar R, Rodrigues A L, Bragança L and Pinheiro D 2019 From temporary to permanent; a circular approach for post-disaster housing reconstruction *IOP Conf. Ser.: Earth Environ. Sci.* **225** 012032

Askar R, Bragança L and Gervásio H 2021 Adaptability of buildings: a critical review on the concept evolution *Appl. Sci.* **11** 4483

Brand S 1994 *How Buildings Learn: What Happens After They're Built* (London, UK: Penguin Books)

Duffy F 1990 Measuring building performance *Facilities* **8** 17–20

Geldermans R J 2016 Design for change and circularity—accommodating circular material and product flows in construction *Energy Procedia* **96** 301–11

Habraken N J 1972 *Supports: An Alternative to Mass Housing* (Westport, CT, USA: Praeger Publishers)

Leaman A, Bordass B and Cassels S 1998 Flexibility and Adaptability in Buildings: The 'killer' Variables. https://www.usablebuildings.co.uk/UsableBuildings/Unprotected/Flexibility1p2.pdf

Meerow S, Newell J P and Stults M 2016 Defining urban resilience: a review *Landscape Urban Plan.* **147** 38–49

Ness D A and Xing K 2017 Toward a resource-efficient built environment: a literature review and conceptual model *J. Ind. Ecol.* **21** 572–92

Schmidt R, Deamer J and Austin S 2011 Understanding adaptability through layer dependencies *Proc. of the 18th Int. Conf. on Engineering Design (ICED 11), Impacting Society through Engineering Design* 5 13

Schmidt R, Eguchi T, Austin S and Gibb A 2010 What is the meaning of adaptability in the building industry? *Proc. of the 16th Int. Conf. on Open and Sustainable Buildings* 227–36

Sinclair B R, Mousazadeh S and Safarzadeh G 2012 Agility, adaptability + appropriateness: conceiving, crafting and constructing an architecture of the 21st century *ARCC J. Architect. Res.* **9** 35–43

**IOP** Publishing

Sustainable Urban Development
Topics, trends and solutions
**Luís Bragança, Cristina Engel de Alvarez and Luisa F Cabeza**

# Chapter 11

# Information technology and the digital transformation of building design

**José Pedro Carvalho, Luís Bragança and Ricardo Mateus**

As society is continuously embracing information technology (IT) in different fields of everyday life, the construction industry will also be affected by this transformation. With the recent deployment of IT in the construction industry, new opportunities and approaches are being sought to enhance the efficiency, sustainability, and general quality of buildings. Among the most recognized forms of IT, building information modelling (BIM) stands out as a major industry change driver. By allowing multidisciplinary information to be overlaid in a single model, BIM creates an opportunity to incorporate sustainable measures in a project. BIM can act through the whole of the building lifecycle, giving it added value for the customer and society. Digitalization impacts will benefit the whole of society in the form of better constructions that consume fewer resources and operate more efficiently. This chapter identifies the current IT applications used in the construction industry, with a focus on the BIM method. Current applications, benefits, and challenges are identified, as well as software and procedures. Moreover, we analyse the integration of digital technologies to enhance construction sustainability, highlighting the impact on project development and quality. Despite the increased use and benefits of IT, current tools and working methods are not yet properly oriented for sustainability purposes. Nevertheless, IT has great potential for sustainable construction, and it is an essential tool for developing high-performance buildings. The main benefits are obtained during the early project phases, either by the provision of reliable data for decision-making or by increased process efficiency.

## 11.1 Introduction

With improving living standards and economic and population growth, societies are increasingly concerned about the demand of building occupants for comfort. To meet this request, society would have to increase the exploration and depletion of

doi:10.1088/978-0-7503-3971-1ch11

natural resources, increasing energy use and social inequality. The effort and cost required to obtain and distribute those resources also create a massive environmental impact. The relation between buildings and environmental impacts has already been proved and accepted by the scientific community. The built environment is also a place for social interaction and culture creation, which often produce several impacts: resource consumption, atmospheric emissions, water consumption, land use, acidification, toxicity, and waste creation. In Europe, buildings are responsible for 36% of the energy consumption and 39% of the $CO_2$ emissions. Facing this scenario, authors argue that the construction industry must increase its efforts to improve social, economic, and environmental sustainability. Also, global governments and societies have been looking for more efficient and sustainable buildings and construction practices, making them increasingly significant over the years. Achieving low-energy buildings is of growing interest worldwide, and it is on the United Nations agenda for 2030. To be aware of this problem, companies need to be more flexible so that they can quickly adapt and respond to market changes, with an urgent requirement for sustainability.

As a result of the increasing environmental awareness, significant interest has arisen in the role of IT. It is known that IT has the potential to address the negative environmental impacts of the building industry; its relationship with ecological sustainability has been attracting growing attention. IT can be seen as a solution to achieving sustainability targets. For companies, IT has already become an important source of competitive advantage in the market. By focusing on sustainability, organizations can improve general operation, innovation, and strategic growth, while gaining a sustained competitive advantage and delivering more sustainable values. Within the architectural, engineering, and construction (AEC) industry, IT has been rapidly adopted. However, the traditional fragmented processes and lack of investment have resulted in poor performance and inefficiencies, mainly due to limitations on information exchange and communication. As designers usually use different and heterogeneous systems, it is hard to achieve full interoperability among all stakeholders. Nevertheless, IT has the potential not only to easily share information within a single organization but also across organizations, improving organizational sustainability and asset management. The benefits of IT overcome the limitations by providing mechanisms for accountability, control, and mutual and reciprocal knowledge sharing, as well as supporting efficient data integration, sharing, and management. Furthermore, IT has been recognized as one of the main approaches for achieving more sustainable buildings and cities.

Among the best-known IT applications for the construction industry, computer-aided design (CAD) clearly stands out. In recent years, BIM has slowly transformed the traditional building design, construction, and communication processes (J. K. W. Wong and Zhou 2015). BIM can be seen as an extension of CAD, as it enables the use of parametric object-oriented modelling, as well as the storage and management of different types of information. The impact of BIM will be felt beyond the construction sector, reaching the whole of society in the form of better constructions which consume fewer resources and operate more efficiently. BIM can be used through the whole of the building lifecycle and also improves communication between all the stakeholders. A standard exchange model, Industry Foundation Classes (IFCs), has been defined and registered by ISO as an official International Standard. The IFC model specification is a '*platform-neutral, object-based structured file format*' that describes building and construction industry data (El-Diraby *et al* 2017). Given the capabilities of BIM, a significant research effort has been made to apply it to building sustainability. Several authors have recognized that BIM capabilities can have a positive impact on building sustainability. By allowing the storage of multidisciplinary data, BIM creates an opportunity to integrate sustainable measures from the project design phase. Additionally, BIM can also enhance the adoption of sustainability practices by developing new tools and plugins. Overall, BIM has considerable potential to enhance the efficiency, sustainability, and effectiveness of buildings. Designers can introduce and compare the performance of different solutions before construction takes place, considering the technological, economic, social, and environmental aspects (Carvalho *et al* 2019). Despite the positive impact of IT and BIM on construction sustainability, relationships still require further research and knowledge. There is a lack of common understanding between researchers, and methods are not properly oriented to building sustainability.

## 11.2 Information technology for building sustainability

### 11.2.1 IT in the construction industry

It has been acknowledged that technology and society coevolve. Recently, several studies have examined the impacts of IT in the construction industry (Kang *et al* 2013). A common conclusion is that IT can have a positive impact on AEC companies, influencing project performance in terms of schedule, cost, quality, process, customer satisfaction, communication, decision support, and safety. Furthermore, it also improves process automation and project sustainability and decreases errors during construction. However, according to (Kang *et al* 2013), IT benefits are inconsistent, leading several studies to reach opposing conclusions. Some authors have effectively noticed advantages in terms of cost and schedule, while others did not find any benefits to cost performance. The possible reasons could be of different types, including measurement errors, time lags in the payoffs of IT, the redistribution and dissipation of profits, the mismanagement of information and technology, sample size and data source issues, or modelling issues.

Within the construction industry, IT can be described as the application of decision support tools (by means of electronic machines and software), which are able to process, store, analyse, control, and transfer construction data during the project lifecycle. The role of IT can be classified into three types (Standing *et al* 2008): automation, information, and transformation automation to replace the human effort; information to provide more data for decision support; and transformation for process restructuring. However, most AEC companies usually focus their attention on process automation as a way of reducing cost and human labour. Regarding the benefits to the AEC industry, they can be divided into: strategic, tactical, and operational benefits. The strategic benefits are related to customer/supplier relations and satisfaction, as well as organisational and process flexibility. The tactical benefits aim to improve the response to changes, improving time plans and cost. The operational benefits relate to data management, communication, decision-making, project quality, forecasting, and control. IT can enhance different project areas by supporting management, enhancing quality and decision-making, optimising resources and efficiency, and providing coordination and communication. Nevertheless, the truth of IT is not derived from the technology itself but is realized by redesigning and improving existing processes and practices. Companies must align IT with their current work practices to deploy information and communications and to increase the benefits of IT. Traditionally, some of the most commonly used IT tools are the internet, email, AutoCAD, and document management.

To minimize construction impacts, the industry has attempted to tackle a wide range of issues using the capabilities of IT. With increasing awareness of sustainability issues, IT can play a key role in construction sustainability and organizational performance. It has the potential to create, enable, and facilitate production and consumption in a sustainable manner, creating a way for construction projects to embrace sustainability. To support building sustainability, IT aims to effectively reduce material and energy consumption as well as waste production during the whole lifecycle. It also allows for simulation, spatial data processing, innovative data exchange, communication methods, networks, data storage, and the implementation of automated technologies; hence, it improves global efficiency. IT can act in all the sustainability dimensions, but worldwide, the focus has been given to environmental aspects (mainly for energy and carbon reduction). However, acting in a more environment-friendly way does not mean acting in a more sustainable way. To do so, the social and economic aspects must be considered. Social aspects are commonly neglected by IT users, while in the economic scenario, the primary focus is always cost. The implementation of sustainable practices lacks a proper integration with IT, where the focus is usually given to environmental sustainability. Nevertheless, researchers agree that IT is highly effective in supporting the sustainable delivery of construction projects.

### 11.2.2 Building information modelling

Recently, BIM has been introducing a paradigm change in the construction industry. As a subset of IT, BIM is a successful method for connecting and automating the fragmented project process, while improving process productivity. BIM can be described as a working methodology that manages the 3D drawing and project data in a digital format during the entire building lifecycle. BIM's essence lies in information sharing and constant communication between all project stakeholders, facilitating decision-making and the identification of errors, while optimising resources. BIM functionalities make it possible to virtually construct the whole building model through simulations and analysis before the construction takes place on site. BIM impacts are noticed throughout the building lifecycle, from visualization, documentation, management, cost planning, maintenance, operation, detailed analysis, and logistics to renovation or demolition. The BIM process implies the development of an object-oriented model which contains all the project data. This BIM model is classified according to the amount and type of data by levels of development (LOD). According to the American Institute of Architects (AIA), the LOD clearly specifies and articulates the model's content and reliability. It ranges from LOD 100—the conceptual model—to LOD 500—as built. Information exchange between stakeholders happens using standardised formats. The most known and used is the IFC. Other formats such as gbXML have also been used, mainly for energy analysis purposes, since they have a higher capability to store thermal data.

Despite all the advantages, it is difficult to adopt new working methods inside companies. This is worsened by the lack of knowledge on the subject, as well as by the initial investment in software and training, which leads to productivity loss during the transition stage. These problems may be worst for small companies, which have a smaller investment capability and fragmented processes. Some interoperability constraints and the lack of normalisation and oriented software are also key factors for companies that are considering the adoption of BIM. To positively influence BIM implementation, efforts must be made starting with the training of architects and engineers. BIM must be a part of curricular plans to endow future professionals with enough knowledge to use and deal with new IT. A proper and clear communication plan can also support BIM divulgation among current construction stakeholders.

Nevertheless, by allowing multidisciplinary information to be overlaid and grouped in a single model, BIM creates an excellent opportunity to introduce sustainable measures into a construction project. In 2008, Krygiel and Nies (2008) listed seven aspects in which BIM could contribute to building sustainability: building orientation, building massing, daylight analyses, water harvesting, energy modelling, and sustainable materials. In fact, BIM can act in several fields for building sustainability, such as performing thermal analysis, identifying solar patterns, estimating water use, managing resources and waste, improving project quality and efficiency, enhancing information sharing, and reducing time and costs. Besides the environmental issues, BIM also impacts other sustainability domains such as the society and the economy. From a social perspective, BIM will support the creation of a healthy and livable community by providing tools to improve building operation, and as an instrument of social engagement by involving occupants. In the economy, BIM will help in assuring economic viability, increasing productivity, and reducing waste. A common agreement is that BIM applicability can be extended to the building lifecycle to support sustainable operation, repair, or demolition. With BIM, designers are able to compare the performance of different sustainable solutions during the early design stages, allowing for the efficient development of high-performance buildings. Furthermore, the design stage has been identified as the critical one in which sustainability-related decisions must be taken. As this is also one of the main stages in which BIM can act, its influence on building sustainability is clear (Carvalho *et al* 2019). J. K. W. Wong and Zhou (2015) have stated that BIM can be used as '*a model-based process of generating and managing coordinated and consistent building data during its project lifecycle that enhance building energy-efficiency performance and facilitate the accomplishment of established sustainability goals*'. Moreover, BIM and sustainability aim at efficient proposals for new architecture and engineering forms, preserving the natural environment and ecosystems. Both concepts intend to achieve a new lifestyle, a new environmental education, and a new perspective for the impact of human development on the planet. With the evolution of BIM, it is expected that building design, construction, maintenance, and decommissioning will be performed in a more sustainable manner. BIM can also facilitate the implementation of LEAN construction techniques, as they are synergistic practices. Some of the LEAN philosophies that can be addressed with BIM are resource optimisation, flexibility towards change, product quality, cost reduction, and mass production.

On the other hand, another common conclusion of researchers is that BIM is still not oriented to sustainable buildings (Carvalho *et al* 2019; El-Diraby *et al* 2017). There is a need to extend BIM use for sustainability purposes, to exploit its environmental benefits and embrace all the building lifecycle stages. There is a lack of software for assessing building sustainability (more sustainability aspects need to be considered) and interoperability errors are still a common experience. As for its other uses, BIM use for building sustainability is also not still a current practice in the industry, mainly due to the lack of knowledge about BIM and its capabilities. To identify and establish the current understanding of BIM sustainability impacts, the following sections present a review of the sustainable fields where BIM can act.

#### *11.2.2.1 Massing*

BIM allows designers to reduce the mass of the buildings by optimising the building shape and envelope. By considering other sustainable analyses, such as wind analysis, project teams are able to enhance the design of their buildings and reduce the required material quantities. A reliable wind analysis can easily be made in Autodesk Green Building Studio (GBS)—figure 11.1—merely by defining the building's location and orientation. Other tools such as Autodesk Flow Design can also provide reliable wind analysis. Computational fluid dynamics (CFD) calculations can also improve building design quality by integrating structural design and wind engineering analyses. Similar proposals have also been made,

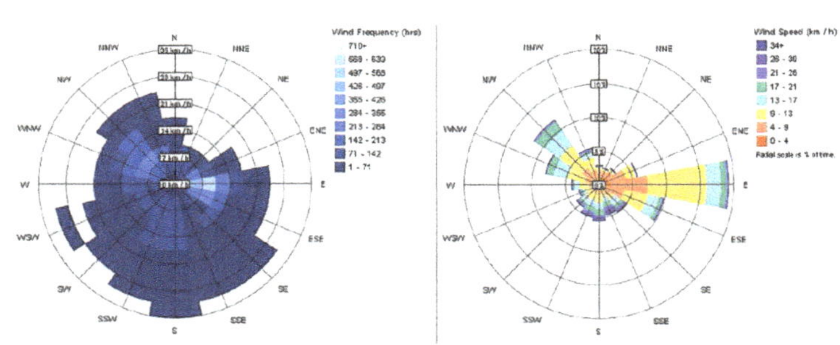


**Figure 11.1.** Wind analysis in GBS.

suggesting a BIM-based CFD simulation to obtain an optimal building design alternative from the wind environment perspective. Mass analysis can also be combined with energy analysis in BIM, helping designers to improve a building's thermal inertia, and therefore reducing the building's energy needs. With respect to wind/mass analyses, typically an LOD of 100 for the building and the surroundings is enough to provide reliable results. An indirect approach to reducing the building mass involves proper project communication and coordination, often avoiding surplus materials, as project quantities are rigorously determined.

*11.2.2.2 Daylight and lighting*

Daylight analysis is among the most common BIM applications for sustainability. Its use has the potential to correctly place the building on the site, as well as define the best window size, location, and shading devices. Solar gains can be optimised during the winter and avoided during the summer. A proper daylight analysis will also contribute to reducing the need for artificial light. Laser scanning technology can generate 3D models to be subjected to lighting analysis, according to a set of requirements. The Autodesk Revit API allows the development of different prototypes to integrate the BIM method and lighting analysis. The date required to perform light analysis in DAYSIM and Radiance can be collected with high efficiency and accuracy. Although BIM does not have all the necessary information for Radiance and DAYSIM, it provides options to incorporate the required information. Recently, a BIM-parametric renovation of a South Korean apartment was made to improve an aged building's indoor visual comfort. This framework enhanced the building envelope system by increasing the daylight factor by 15%, daylight autonomy by 30%, and useful daylight illuminance by 15%. Within BIM, a natural lighting analysis can be included in an energy-saving design, a green buildings for energy-saving design, and a lighting simulation analysis. Some other software packages that also carry out daylight analysis (figure 11.2) include DesignBuilder and Integrated Environmental Solutions—Virtual Environment (IESVE).

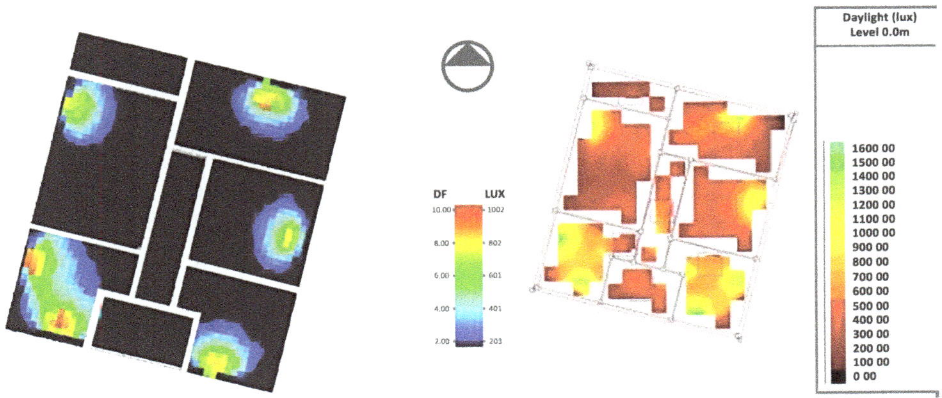


**Figure 11.2.** Daylight analysis in GBS and IESVE.

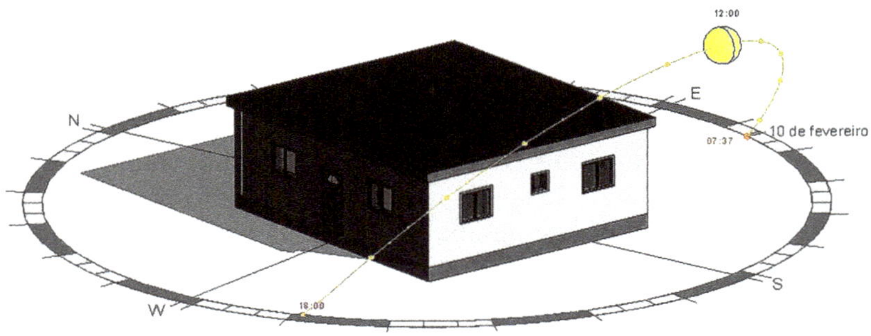


**Figure 11.3.** Building orientation analysis in Autodesk Revit.

### *11.2.2.3 Building orientation*

Building orientation analysis is often combined with wind and daylight analysis. As presented above, wind analysis can substantially improve building design in terms of mass. Additionally, wind analysis is also a valuable aid in defining window location and size to promote natural ventilation. An appropriate orientation selection can have considerable impacts on the occupants' comfort, as well as on the energy bill. Building orientation can have a considerable impact on a building's energy consumption, with cumulative savings of around €1000 for a 30-year lifetime period. The impact of windows positioned in different orientations can also be analysed by simulating different scenarios in GBS, achieving annual energy consumption variations of 1% for a small-scale dwelling. Nowadays, BIM software such as Autodesk Revit and IESVE allows the user to perform shading analysis by day and hour, offering a deep understanding of the impact of building orientation (figure 11.3). When an energy analysis is made in GBS, the software automatically performs an analysis of the building energy consumption for different building orientations. Designers can compare the orientation impacts on the building's annual energy consumption.

### *11.2.2.4 Energy modelling*

Energy performance analysis is the most common BIM analysis nowadays, in combination with building energy modelling (BEM). Given the ease of performing an analysis of the potential benefits of different design approaches, designers commonly use them from the earlier design phases in order to enhance building energy performance and thermal comfort.

A BIM-based API was already developed to create Modelica energy models. Attributes from Autodesk Revit models, such as building topology, geometry, and materials were converted into parameters for Modelica and used in energy simulations. Energy efficiency analyses can also be made using EnergyPlus. An industrial building has been submitted to a set of simulations, producing high-impact savings. However, existing procedures are still not mature enough, requiring a significant amount of time, assumptions, and remodelling. Efforts have been made to assess the energy and atmosphere credit defined by the Leadership in Energy and Environmental Design (LEED) method through a simulation process based on

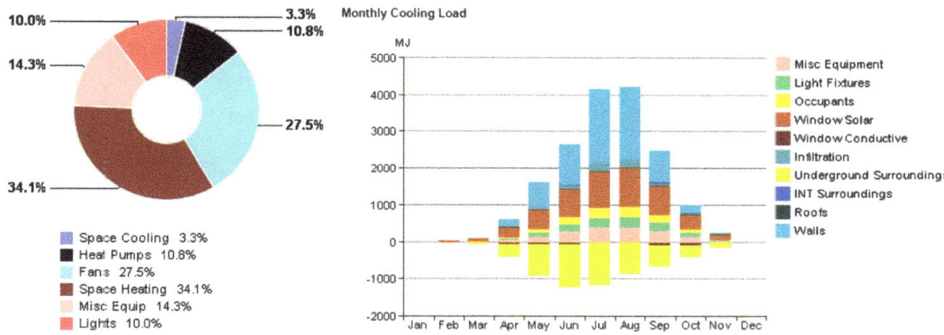


**Figure 11.4.** Graphical results from GBS.

Autodesk Revit. The model geometry was easy to validate, and BIM was able to make a significant impact in building energy simulations. However, considerable time was wasted in double-checking and correcting geometry errors. This same problem had already been identified by various authors, who mainly referred to interoperability problems with the model's geometry. Among the main benefits of BIM, we should mention the decision support provided to designers in the early project stages. They are able to compare the impact of different construction solutions in a short time period using few resources.

Regarding software, IESVE and GBS are the most commonly used products, as IESVE is the most versatile and powerful, while GBS provide fast, reliable, and user-friendly energy analyses (figure 11.4). GBS and IESVE have been jointly recognized as the most all-embracing software packages for green analyses. Both can perform energy, carbon emission, natural ventilation, solar, daylight, and water analyses, and both lack the ability to perform acoustic analysis. Concerning the BIM model's LOD, authors have found that an LOD of 300 is required for a proper energy assessment. The model must contain accurate data about quantities, size, location, orientation, and material properties.

### *11.2.2.5 Water*

The efficient use of water in buildings is also a field where BIM can have a positive impact. Currently, several BIM software packages allow the prediction of operational water consumption and the integration of sustainable measures. GBS is one of them, and by importing a gbXML file from Autodesk Revit, the amount of water required for a building can be estimated. A harvesting system or other net-zero measures can also be introduced to increase water savings. The role of BIM in improving the design process for water supply and drainage in buildings, and therefore, building quality, has already been highlighted. It can also support efficient rainwater harvesting and reuse, as well as provide real-time monitoring of water use. BIM can also be used at the city scale by integrating a geographic information system (GIS) method for the water distribution system planning. This method can effectively identify conflicts between the water project and existing structures, pointing out any existing conflicts before construction takes place. By linking the analysis results with LEED, Azhar *et al* (2011) have assessed two credits in the 'water efficiency' category and one in the 'innovation in design' category by using IESVE water simulation capabilities. The recommended water-reduction measures achieved a 47% reduction.

*11.2.2.6 Acoustic*

Acoustic analysis is usually performed for specific and special buildings, such as theatres, auditoriums, and concert halls. Due to the complexity of the analysis, this type of process is only applied to significant buildings in which the acoustics are a key role in the public experience. With the introduction of BIM, the possibility arises that designers can simulate the acoustics of buildings, as much of the required information for the analysis can be assessed through the data contained in the BIM models. Several software prototypes have been created using the Autodesk Revit API so that designers can quickly perform acoustic simulations. By creating custom parameter schedules, prototypes are able to extract the data required to perform an acoustic analysis of rooms for reverberation and sound intensity levels. Such schemes provide designers with a decision support tool that allows them to achieve better building designs. On similar lines, the construction software development company Cype has created a specific software package for Portuguese buildings—Cypesound RRAE—that allows designers to quickly perform acoustic simulations according to the actual Portuguese regulations for the building's acoustic requirements. This tool can be populated with data using the Open BIM process, as Cype has already released several tools to adapt IFC files to their software. Cype also has available other region-oriented software, such as Cypesound CTE for Spain and AcouBAT for countries such as Brazil, France, or Italy. Usually, authors tend to develop a BIM-based API to simulate the acoustics of concert halls (e.g. using Comsol software). They claim that the simulation time has been reduced and they had the opportunity to compare the performances of different design scenarios. However, they still faced some limitations on the building geometry and materials.

*11.2.2.7 Sustainability assessment*

The use of BIM to support building sustainability assessment methods has been a growing topic among researchers in recent years. The assessment of this kind of method exploits the full potential of BIM for building sustainability, as multi-disciplinary data must be assessed using different BIM analyses and software. Traditionally, the LEED method has been the method most selected for BIM-related research. Azhar *et al* (2011) validated a conceptual structure to connect BIM and LEED using Autodesk Revit and IESVE software; they gathered the required data to assess 17 out of 29 LEED criteria and two out of three prerequisites. Other authors have also focused attention on LEED, assessing 'energy and atmosphere' and 'material and resources' credits from LEED v3 using only Autodesk Revit. Autodesk Revit has itself been identified as the software most used for this purpose by researchers. J. K.-W. Wong and Kuan (2014) assessed 26 out of 56 criteria from BEAM Plus; BIM provided a faster assessment with fewer resources. Sixty-six percent of the credits of the Australian Green Star Building Certification were been assessed. Even so, they stressed that the BIM capabilities were not used to their full potential and that BIM execution and coordination plans must be developed for green building certification. They also noticed the importance of creating proper guidelines for BIM integration. Carvalho *et al* (2019) suggested a conceptual framework for a BIM-based assessment for the Portuguese method SBToolPT-H. Furthermore, they also analysed the attraction of further developing the framework into an API. Based on current research, several authors have proposed that future BIM standards should include requirements for building sustainability assessment.

#### *11.2.2.8 Site analysis and waste management*

BIM can be used from the beginning of the project design phase, by performing a set of site analyses, in which contractors can analyse the site conditions, such as wetlands or protected habitats, as well as efficiently plan the site utilisation. Furthermore, BIM can support all the construction-phase logistics, coordination, and planning, as well as the maintenance schedule, assets, and space management during the operational phase. The benefits of BIM are also noticeable in the demolition and renovation phases with regards to waste management. As an example, a BIM system prototype was created and applied (Autodesk Revit API-based) for the estimation and planning of demolition and renovation waste. The system was able to extract material and volume information through the BIM model, as well as consider waste recycling and reuse, pick-up truck requirements, and waste disposal fees. A computer algorithm can automatically calculate the quantities for a design option so that BIM can instantaneously map them to the waste-generation-rate database and estimate the potential amount of waste. This kind of system can be useful to support designers and contractors in managing construction waste. BIM can also support construction and demolition waste managers in estimating the amount of generated waste, as well as inventory and equipment location and size. Furthermore, it can also support the identification of waste area and volume, which are required to effectively calculate the number of pick-up trucks and bins required. Waste cost planning and tracking can also be supported by BIM. However, none of the existing construction and demolition waste management tools is BIM compatible. Nevertheless, BIM can support contractors in the planning processes, reducing waste and carbon footprints.

#### *11.2.2.9 Sustainable materials (LCA)*

BIM is widely used for the selection of sustainable materials; this use is still an emerging trend. The combination of BIM and lifecycle analysis (LCA) allows designers to understand and compare the impacts of their materials. Soust-Verdaguer *et al* (2017) stated that the integration of BIM and LCA can reduce the effort required for data input and can also be applied to the comparison of building materials and energy scenarios. BIM will provide the required data for stakeholders to perform LCA, which they usually do not have in the early design stages. BIM can be linked to LCA in three different ways: through the life cycle inventory (LCI); in addition to LCI, environmental information can be integrated into BIM software; and automated processes combining different data and software. A combination of Autodesk Revit, GBS, and Tally has also been made, creating the possibility of evaluating several environmental impacts of a building, from the material manufacture to the end-of-life scenario. BIM–LCA integration has been suggested as an essential step for achieving a more sustainable built environment and empowering the decision-making process. Hollberg *et al* (2020) used Autodesk Revit and Dynamo to assess the embodied global warming potential (GWP) in the post-design assessment phase. The results showed that BIM can reduce the effort of calculating those impacts during the design phase. However, they call attention to the BIM model quality and selected tools, which can lead to wrong results. LCA must be performed from the early design stages. During the conceptual stage, the BIM model should have an LOD of 100, while during the design stage it should have an LOD of 300. Soust-Verdaguer *et al* (2017) agree that an LOD of 300 is the most appropriate for verifying the environmental impacts during the early design stages. Other studies have extended the use of BIM–LCA results to assess a sustainability scheme. Using a Tally Life Cycle Inventory, the data required to assess one criterion of SBToolPT-H were automatically gathered. Since LCA usually neglects social and cultural aspects, this should be performed using other multicriteria assessment tools to increase the usefulness of LCA. Thus, Hollberg *et al* (2020) suggested that LCA and sustainability certification should develop a common database for long-term use.

## 11.3 Other innovative approaches

Some other IT approaches are being sought to enhance the performance of buildings and construction. Among some of the innovative approaches, the concept of a smart city (and buildings) stands out as an approach that supports innovation and economic growth, as well as enhancing sustainability, efficient urban management, and development (Jamei *et al* 2017). Smart cities are incorporated in city planning using IT-based services for urban development plans. The aim is to increase the efficiency and betterment of infrastructure (for energy, waste, or transportation) and people (in the form of safety, knowledge, or participation). Different definitions have been given for this concept. Overall, smart cities intend to improve life quality and achieve sustainability and efficiency goals by investing in human and social capital, as well as in traditional and modern communications infrastructure, with a wise management of natural resources, through participatory governance.

The development of smart sustainable cities is usually based on the Internet of Things (IoT) and related to big data analytics. IoT, as a computing paradigm which is related to the big data analytics trend, is stimulating smart sustainable city initiatives, as a path for optimizing energy efficiency and environmental impacts across many urban domains. The IoT defines a computationally augmented environment in which the physical and the informational worlds are integrated within the internet infrastructure via a wide range of active and smart data-sensing devices. The IoT offers the capability to integrate digital and physical entities, enabling a whole new class of big data applications and services that can enhance environmental sustainability. As expected, sensor technology and big data are some of the cores enabling the IoT. Sensors are classified according to the types of energy they can detect as signals (such as optical/visual, light, image, sound, temperature, heat, pressure, motion, physical, etc.), and used to collect large amounts of data that serve as inputs for big data applications. Thus, the large amount of data collected by sensors is turned into useful knowledge for decision-making and enhanced insights related to urban domains, such as transport, mobility, traffic, environment, energy, land use, planning, and design. The concept of big data can be summarized as the processing and analysis of: volume, variety, velocity, veracity, and value to create actionable insights for sustained value delivery, measuring performance and establishing competitive advantage. It is often associated with social and environmental performance, especially when linked to organizational and coordination capabilities within supply chains. Overall, a direct connection can be made between smart sustainable cities, IoT, smart sensors, and big data. The IoT deals with countless physical and virtual objects and involves a set of sensors. However, it generates a large volume of data, and the processes and platforms required to manage these data are complex—big data has emerged to optimise the storage, integration, processing, analysis, and management of such data through scalable applications. The IoT, together with such IT novelties, has demonstrated the required potential to add a whole new dimension to the application and enhancement of environmental sustainability across several urban domains of smart sustainable cities (Bibri 2018).

With a differentiated approach, Jamei *et al* (2017) analysed the impact of virtual reality (VR) on smart cities. Despite the lack of related research, VR can offer key benefits to smart cities, such as assessing design ideas with 3D space in real time during the planning phase, the integration of different design aspects, and the promotion of social and participatory planning. VR is an interactive and immersive 3D virtual environment, where users control the viewpoint and realistically interact with various components. VR can realistically simulate scenarios of acrophobia, disability, social psychology, and emergency evacuation and also promote stakeholder communication in an immersive environment. Thus, VR has been attracting increasing attention from the AEC industry. VR is commonly used in the industry for design reviews and coordination, lighting design, planning, operation, education, maintenance, facility management, and safety training. With the development of new IT, such as BIM and GIS, VR is expected to continuously draw attention and will be widely used in the AEC industry. The real-time analysis of different scenarios will help governments, stakeholders, and communities to be aware of the environmental and social impacts of their projects.

Despite its popular use since the beginning of 2000 for the digitalization of machinery and production systems, 'digital twins' is one of the last terms to have emerged from IT into the construction industry. The concept can be described as a digital mirror image or a representation of a physical process or object, which includes the integration of assets, processes, or systems for virtual simulations. Within the sustainability field, digital twins can be related to circular economy principles, allowing for constant access and updates to the digital representation of physical assets. By enabling the transmission of data between the physical and virtual worlds, digital twins can facilitate monitoring tasks and optimize the functions of all the physical entities. Digital twins have been used in the construction industry for different applications, such as road construction and smart buildings. With the integration of IoT, data sensors, and BIM, the digital twins concept has emerged, allowing for constant monitoring of the physical environment and consolidation of enormous amounts of data for real-time decision support. Despite its nascent status, digital twins can significantly enhance process management, productivity, and efficiency.

In another field, laser scanning is also an emerging technique. Recent studies have applied a laser scan to carry out dimensional surveys and digitalization of existing buildings. For instance, the state of conservation of an Italian historic monument can be assessed. With laser scanning, significant time and costs were saved when analysing the actions required to preserve the architectural heritage and culture. Another example of an IT application is the use of terrestrial laser scanning (TLS) for moisture detection in buildings. With TLS, building saturation and reflection can be assessed to improve construction quality and comfort. TLS can also be used in combination with BIM (Scan to BIM) by creating BIM models based on TLS data. Furthermore, TLS has been identified as an appropriate technology for collecting the 3D data of a project with high accuracy and speed.

Another IT innovation for the construction industry concerns 3D printing, which is also an emerging topic. In 3D printing, physical objects are created by depositing materials in layers based on a virtual model. Therefore, its connection with BIM is very substantial, as BIM can provide the 3D printing software with information that governs the design and construction process. The benefits for the construction industry start with lower costs for printed construction elements, material transport, and storage. From an environmental perspective, it can support the use of low embodied raw materials and reduce the waste produced by construction processes. It can also save considerable time during construction, compared to traditional processes. Overall, it is believed that 3D printing will increase building sustainability by selecting materials based on lifecycle analysis and by supplementing traditional construction techniques.

## 11.4 Sustainable buildings

It is known that the use of BIM can provide significant benefits to building performance and also save considerable resources. The use of BIM to develop sustainable buildings can provide a 20% saving on the total cost of a project in a ten-year period. As stated earlier, the Shanghai Tower (figure 11.5) is a successful building where BIM was applied to enhance building sustainability. Facing complex design and construction challenges, there was a need to improve traditional methods to properly manage and control the project, as well as to achieve the sustainability targets. One of the most recognised pieces of evidence about the building was the 32% reduction of the materials required for the same building area (15% of which was related to the façade glass). Additionally, by using BIM to perform wind analysis, the building shape was optimised to reduce wind loads by

**Figure 11.5.** Shanghai Tower, China (Balye, 2019).

24% and, therefore, the amount of steel required for the building's structure, creating a saving of 12 million dollars per 5% of steel removed. In total, the 121-storey tower has 43 different sustainable technologies, including renewable energy sources, rainwater collection, water reuse, extensive landscaping, and an optimised shape. Together, these technologies have provided a 21% reduction in the building's energy consumption. The carbon emissions were also reduced by 37 000 tons per year. According to the Shanghai Tower Director and General Manager, BIM technology has supported the project team in producing a high-quality project, avoiding on-site changes which would have required time, material, and human resources.

## 11.5 Research projects

The European Union (EU) has recently funded some research projects linked to BIM and building sustainability. The Buildings As Material Banks (BAMB) project aimed to enable a systemic shift in the building sector by creating circular solutions and ways to increase the value of construction materials through the incorporation of dynamic and flexible buildings in circular chains. The goal was that, at the end of a building's lifecycle, instead of materials becoming waste, they could be reused and/or recycled. The tools developed by BAMB (such as material passports) have gained greater usability as a result of BIM integration (mainly due to the BIM data storage capacity) and enabled materials, components, products, systems, and buildings to be retrieved, reused, and upgraded. A Material Passport (MP) can be defined as the qualitative and quantitative documentation of the material composition of a building. It contains data about the building materials, including their recycling potential and environmental impact. The BIM-based MP can be seen as an optimization tool in the early design stages and acts as an inventory at the end of the lifecycle of a building, serving as the basis for a secondary raw material stock inventory. It enables an optimization of the building design for resource use, recycling, and environmental impacts. BIM is increasingly used in optimizing the management operation of a building during its lifecycle. Information about materials and components is accessible, which is essential to optimize their reuse, reducing the investment in new products.

Another EU funded project is BIMcert, whose objective is to develop a blended, fully supported suite of BIM learning tools to use technology to enhance information exchange and collaboration. The project focus is on testing BIM approaches to green and passive building design to contribute to the improvement of building energy efficiency. BIMcert combines professional skills for energy efficiency with a modern construction method.

More recently, the Circular EcoBIM/SECClasS project has been funded by EEA Grants in Portugal through the Iceland, Liechtenstein and Norway grants program; it aims to create to create a BIM standard product data template for BIM objects. A set of applications will be developed to produce circularity passports for buildings and components, considering the integration and development of Environmental Product Declarations (EPDs) and LEVEL system indicators for buildings.

## 11.6 Conclusions

The deployment of IT usually replaces old problems with new ones. With the increasing use of IT, there is an increasing requirement for industry-oriented software and better interoperability. The financial investment required and a lack of vision and knowledge are the major barriers that prevent companies from successfully adopting IT and gaining a competitive advantage. Nevertheless, the IT benefits for building sustainability are vast, for example, the productivity and organizational improvements, easier communication and collaboration, better support for decision-making, or the development of high-performance buildings. However, IT does not only benefit sustainability, as there is often a time lag before benefits are achieved. Simultaneously, IT use has contributed to increased power consumption, the overuse of scarce resources, waste production, and toxic emissions.

Among the most known and used forms of IT for building sustainability, BIM stands out as a major driver of change. Overall, it has been found that BIM can have significant impacts on improving building sustainability and project efficiency. Many of the benefits are most noticeable during the project phase, in which they take the forms of decision support and resource savings. Moreover, the construction and operation phases can also benefit from efficient management and the coordination of assets and resources. BIM can also support the demolition or renovation stages by managing and quantifying waste. Still, there are some existing barriers, both in the application of BIM for sustainability purposes and in the general adoption of BIM by construction companies. Software companies must start to incorporate sustainability issues into their products and also improve their interoperability capabilities. In the existing frameworks, it is common for information to be lost or to undergo inaccurate data transfer, requiring additional time to check and correct the models.

Other innovative IT solutions that can support building sustainability have also been identified, such as VR, 3D printing, and laser scanning. However, none of the existing IT solutions was developed to directly support building sustainability. Tools and software are not properly oriented towards sustainable buildings, generating a knowledge gap in this field. Besides the commitment required from software development companies with sustainability objectives, designers must also be aware of both sustainability and IT issues. Architectural and engineering teaching courses should update their programs to integrate more comprehensive IT solutions, such as BIM and the required multidisciplinary knowledge. Future designers must be familiar with such techniques in order to facilitate and improve their use in the construction sector. By providing knowledge to stakeholders and given the constant innovation and demand for sustainable buildings, it is expected that IT can play a key role in enhancing the sustainability of buildings and construction.

## Acknowledgments

The authors acknowledge the Portuguese Foundation for Science and Technology for its financial support of this study, through the Regional Operation Programme of the North (Reference SFRH/BD/145735/2019).

## Bibliography

Azhar S, Carlton W A, Olsen D and Ahmad I 2011 Building information modeling for sustainable design and LEED® rating analysis *Autom. Constr.* **20** 217–24

Balye S 2019 Shanghai Tower, Shanghai, China https://unsplash.com/photos/cYne7_kcI80 (Accessed 3 November 2021)

Bibri S E 2018 The IoT for smart sustainable cities of the future: an analytical framework for sensor-based big data applications for environmental sustainability *Sustain. Cities Soc.* **38** 230–53

Carvalho J P, Bragança L and Mateus R 2019 Optimising building sustainability assessment using BIM *Autom. Constr.* **102** 170–82

El-Diraby T, Krijnen T and Papagelis M 2017 BIM-based collaborative design and socio-technical analytics of green buildings *Autom. Constr.* **82** 59–74

Hollberg A, Genova G and Habert G 2020 Evaluation of BIM-based LCA results for building design *Autom. Constr.* **109** 102972

Jamei E, Mortimer M, Seyedmahmoudian M, Horan B and Stojcevski A 2017 Investigating the role of virtual reality in planning for sustainable smart cities *Sustainability* **9** 2006

Kang Y, O'Brien W J and Mulva S P 2013 Value of IT: indirect impact of IT on construction project performance via best practices *Autom. Constr.* **35** 383–96

Krygiel E and Nies B 2008 *Green BIM: Successful Sustainable Design with Building Information Modeling* (New York: Wiley)

Soust-Verdaguer B, Llatas C and García-Martínez A 2017 Critical review of bim-based LCA method to buildings *Energy Build.* **136** 110–20

Standing C, Jackson P, Chen A J W, Boudreau M C and Watson R T 2008 Information systems and ecological sustainability *J. Syst. Inform. Technol.* **10** 186–201

Wong J K-W and Kuan K-L 2014 Implementing 'BEAM Plus' for BIM-based sustainability analysis *Autom. Constr.* **44** 163–75

Wong J K W and Zhou J 2015 Enhancing environmental sustainability over building life cycles through green BIM: a review *Autom. Constr.* **57** 156–65

**IOP** Publishing

Sustainable Urban Development
Topics, trends and solutions
**Luís Bragança, Cristina Engel de Alvarez and Luisa F Cabeza**

# Chapter 12

# Energy efficiency in the change to nearly zero-energy buildings

**Anderson A Fraga, Edna A Nico-Rodrigues, Luisa F Cabeza and Cristina E de Alvarez**

The lack of management of non-renewable resources in the past is the current cause of serious socioeconomic losses, such as energy crises, climate change acceleration, the population's lack of access to resources, etc. World energy consumption now aims to mitigate the impacts caused by the extraction of raw material for electricity production. The indication of a global increase in energy consumption, which shapes the ongoing global energy crises, demonstrates the significance and role of energy-efficient buildings in terms of energy security and renewable energy sources. This study addresses zero-energy buildings (ZEBs) and their efficient use of energy, as well as policies and the applicability of ZEB principles to residential and commercial buildings. Why be energy efficient? And how do government policies approach the subject? It has been observed that climatic, cultural, and construction system particularities are crucial aspects in the energy assessments of buildings, along with public and private incentives for obtaining energy-efficient buildings.

## 12.1 Introduction

The evolution of the human species was largely due to the use and development of technologies that made it possible to overcome the obstacles imposed on its survival. These discoveries necessarily required the use of the available natural resources, and the relationship between their rational consumption and the end product was, until recently, a concern left in the background by society.

However, it is evident that the lack of management of non-renewable natural resources in the past is now causing serious socioeconomic losses, such as energy crises, the acceleration of climate change, and the population's lack of access to resources, especially for the most underprivileged, among other aspects. The population increase in urban centers is a factor that directly influences energy

doi:10.1088/978-0-7503-3971-1ch12  

consumption and carbon dioxide emissions, as can be seen in the graph for the last few decades (figure 12.1).

Considering the period from the beginning to the middle of the 20th century, the US and Europe were responsible for 90% of the total pollutant emissions at the dawn of the twentieth century and an increase of 85% in each subsequent year until 1950. This pattern was also followed by other nations as they adopted the same form of primary energy production and consumption, as shown in figure 12.2.

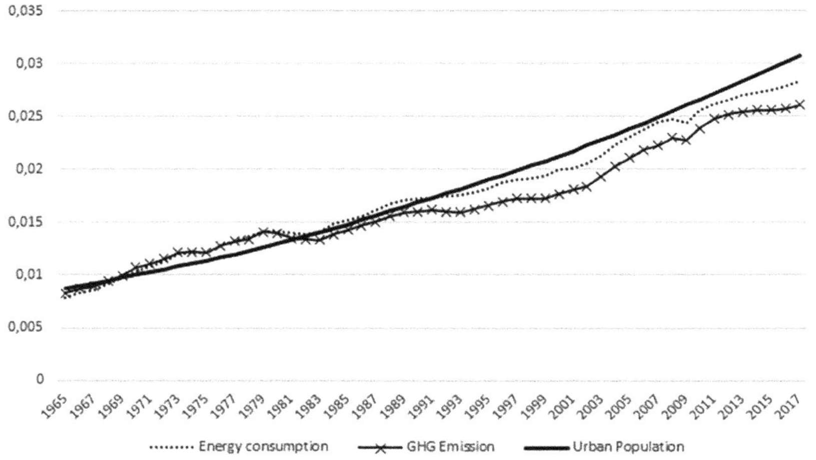


**Figure 12.1.** Primary energy consumption, $CO_2$ emissions, and global urban population growth between 1965 and 2017.

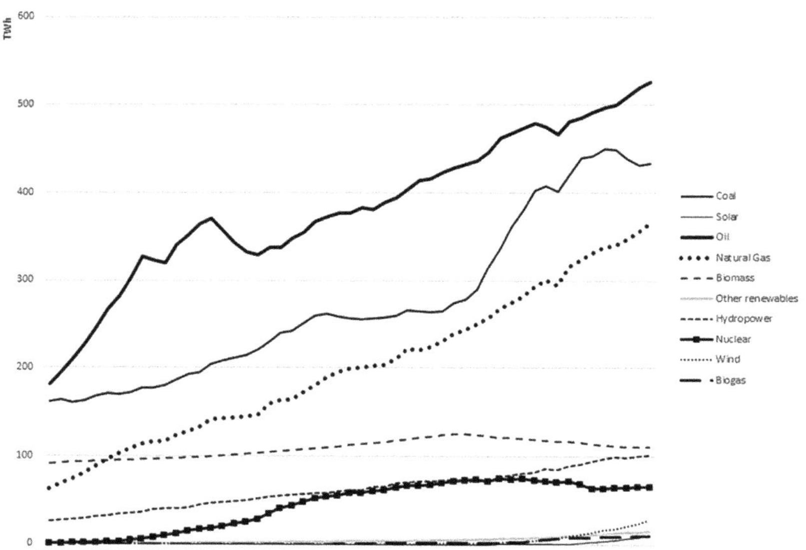


**Figure 12.2.** Global primary energy consumption by source between 1965 and 2017.

World energy consumption paves the way for a reduction of the impacts caused by the extraction of raw materials for the production of electricity; the goals associated with that reduction are constantly reviewed. The extraction process used to produce energy around the globe still depends on non-renewable sources of energy, mostly fossil fuels, which not only causes an increase of pollutant gas emissions, but the also degradation of the environment.

Given the need to lower this impact, policies are regularly agreed between authorities in order to make the relationship between population growth, energy consumption, and pollutant emissions more efficient. In 2018, the absolute contributors to the increase in $CO_2$ emissions were China (28%), the US (15%), the EU (28 member countries, 9%) and India (7%), who produced 57% of the total global $CO_2$ emissions. It is noteworthy that the emissions growth rates of these countries were not directly proportional to their population growth. Out of the total $CO_2$ emissions, one third was produced by China.

It is clear that clean energy sources are becoming more predominant. The data for the energy consumption in Europe from 1965 to 2020 (figure 12.3), combined with the current state of rapidly improving energy sources on the continent, such as solar and wind power, as well as projections for the use of renewable sources, for example, point to the dominance of fossil fuel sources until 2040.

Figure 12.4 shows the installed capacity for different types of renewable energy production over the last 16 years, demonstrating the global commitment to replacing primary fossil energy sources with renewable energy sources.

On the European continent, in some future scenarios, Europe may reduce its primary energy consumption to 29.6% in the least promising scenario and 38.8% in the most favorable scenario, at which point, all the mitigation measures and stipulated targets would be achieved. In addition, the use of renewable energy brings indirect gains not only for the economy and the environment but also for the community, which, in general, benefits from the further development of living

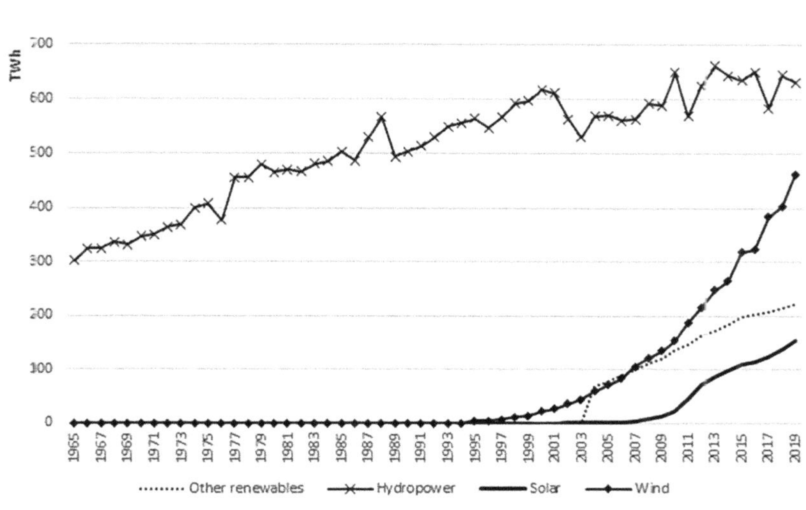


**Figure 12.3.** European consumption of renewable energy by source.

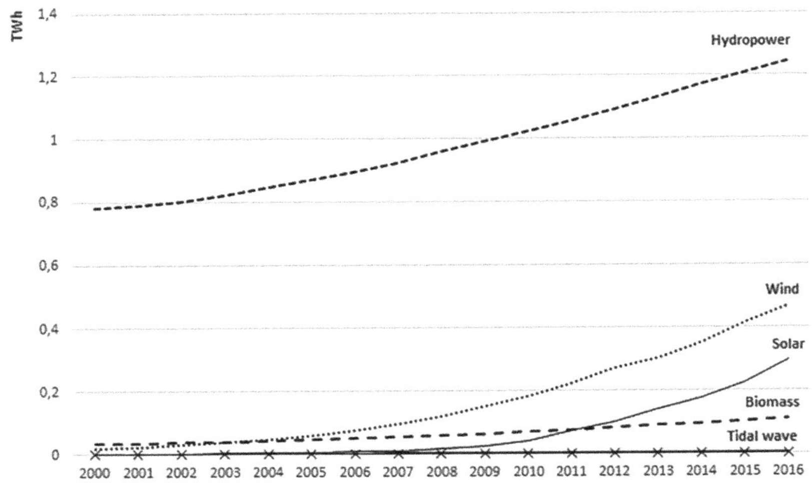


**Figure 12.4.** Worldwide installed renewable energy capacity, by technology.

standards. This development is in non-correlated areas, such as: security, education, employment, cultural, and recreational opportunities. In terms of energy efficiency, the leading questions that guided this chapter were: why be energy efficient? How do programs and government actions address this subject?

The complete adoption of renewable energy as a primary source is a strategy pursued by several global actors; moreover, it is an achievable goal for most countries, despite the required investment and the long execution time.

In the context of climate change, it is a fact that as a result of the increase in energy consumption by cooling systems required to achieve acceptable levels of thermal comfort, the energy drain will jump from 2% to 40% in the most drastic climate scenarios. The authors also emphasize that climate studies are important for the management of buildings, primarily so that energy consumption can be forecast, making it possible to foresee the increase in energy demand. The effectiveness of energy efficiency measures should be considered in the context of regional consumption standards, bearing in mind that policies that aim to reduce environmental impacts need to be adapted to regional characteristics.

Therefore, this chapter addresses the efficient use of energy in ZEBs, considering its applicability to residential and commercial buildings, and also covering programs, regulations, and standards related to the topic.

## 12.2 Energy consumption in buildings

Buildings represent a large part of the total energy consumption in the world. This fact makes the consideration of the building lifecycle essential in order to reduce the impacts on materials and their consequences for inhabited environments. While acknowledging this, there are few data records currently available for material lifecycles, which represents a great barrier to the evaluation of environmental impacts, which also cannot consistently represent the region under review in this chapter.

Ramesh *et al* (2010) found that self-sufficient buildings can retain the embedded lifecycle energy more than some low-energy buildings, due to the energy embedded in the technologies adopted for energy generation. Dascalaki *et al* (2021) demonstrated that this embedded energy can reach 30% more over the building's lifetime, due to the materials used. However, Weißenberger *et al* (2014) highlighted the great potential for interference and the consequent reduction of environmental impacts when considering the evaluation of the lifecycle of materials through the project phases.

Since 2010, research has shown that the use of construction strategies reduces energy consumption in the operational phase and increases the amount of embedded energy, reducing the total lifecycle's energy consumption by approximately half. Other surveys have highlighted that reducing operational consumption increases embedded energy, using an assessment of the adopted strategies versus energy consumption in the operational phase versus embedded energy.

The embedded energy issues, along with climate change during pandemics, are recent topics examined by Brunetta *et al* (2021). They recorded a decrease in the overall energy demand and $CO_2$ emissions in Europe as a consequence of the restrictions imposed as a result of the pandemic. In contrast, Monzón-Chavarrías *et al* (2021) observed a change in daily habits in Spain as a result of the pandemic, which increased energy consumption by 18% along with an increase in the $CO_2$ emitted by residential buildings.

Although it is still important, the demand for electricity has been modified in some regions of the world as a result of the measures imposed, such as lockdowns. The hourly demand for electricity in Spain, for example, offered a glimpse of the modifications imposed on the functioning of the commercial and industrial sectors during the pandemic, showing changes to weekday consumption, which started to resmble weekend consumption. Some examples of important markets for the consumption of electricity in buildings and their current states are as follows:

- In 2011, European countries foresaw a demand for about 25 billion built square meters, with an annual growth projection of 1%. Residential and commercial buildings mainly accounted for this built space, and consumption was expected to be more inclined to warmer spaces, compared to countries with a mild climate, such as Australia and the United States. The part of the European built environment represented by non-residential buildings makes up 25% of the total energy consumed. The energy consumption intensity of these buildings is, on average, 280 kWh $m^{-2}$, 40% higher than for residential buildings. Despite this, in 2020, there was an increase, mainly during the week, of more than 40% in some countries of the European bloc, compared to the same week in 2019;
- Even with the direct effect of the pandemic on global energy demand, the population of the United States accounted for about 53% of the world's primary energy consumption in 2019. In contrast, the construction industry in many US cities and states (such as California, Florida, Idaho, Illinois, Utah, Vermont, and Wisconsin) is increasing the energy efficiency requirements of residential and non-residential buildings by applying energy efficiency codes to buildings;
- In China, the demand from the industrial sector faced a big drop, as in the construction and manufacturing sectors, reducing 69% of the total demand in

2019 by 12%. Residential electricity demand increased, as in practically all economies in the world, as a result of the implementation of lockdown measures;

- In India, the impact suffered in the first quarter of 2020 was less than in other countries, with a 30% increase in demand compared to the same period in 2019. This demand only fell during working days when the country adopted the lockdown at the end of March as a measure to prevent COVID-19 from spreading through the population. In response, the Indian energy ministry launched the Energy Conservation Building Code (ECBC) program, which consisted of an energy conservation code for residential buildings (ECBC-R) with the purpose of improving energy efficiency in the design and construction stages of houses, apartments, and villas. Given the pace of expansion of the Indian civil construction market, the program's main aims were to set up a benchmark and introduce a policy.

There is a concern regarding energy efficiency but also a need to modify the primary energy matrix, emphasizing that the transition to a 100% renewable energy system is a process with several economic and technical challenges.

## 12.3 Zero-energy buildings

Energy-efficient buildings are systems that consume small amounts of energy as a result of either passive or active actions, such as consumption mitigation. A ZEB produces, consumes, and eventually exports energy back to the grid when climatic and energy production conditions are favorable. The evaluation of production versus consumption relies upon the availability of the renewable energy supply, whether it is on site or off site, and the timespan considered for the evaluation process, such as hourly, monthly, or, more commonly, annually.

The ZEB concept originated in the early 1920s and very quickly became popular, but it lacked a consensual definition of the term (Cabeza and Chàfer 2020). Figure 12.5 presents reference strategies for achieving a ZEB. Two strategies are fundamental in order to obtain a building with zero energy usage: the reduction of energy demand and the implementation of renewable energy systems.

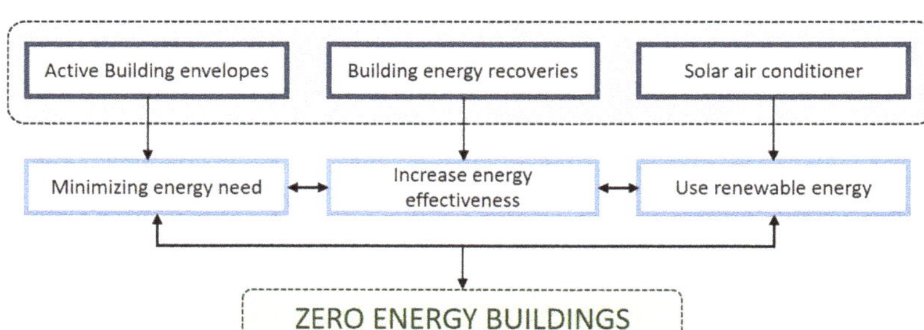


**Figure 12.5.** Strategies for achieving a ZEB building.

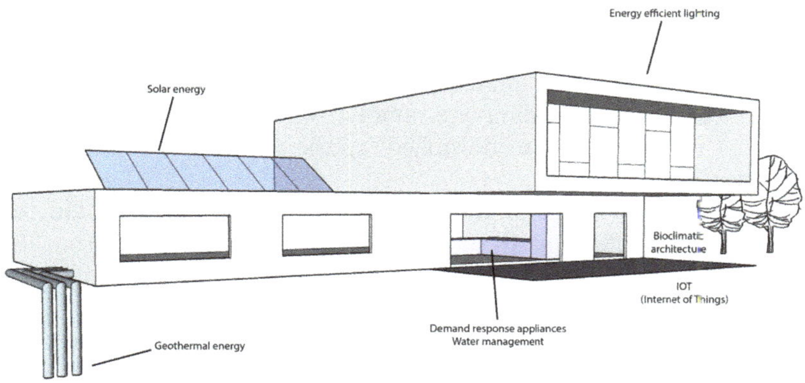


**Figure 12.6.** Building technologies that reduce energy consumption and provide automation.

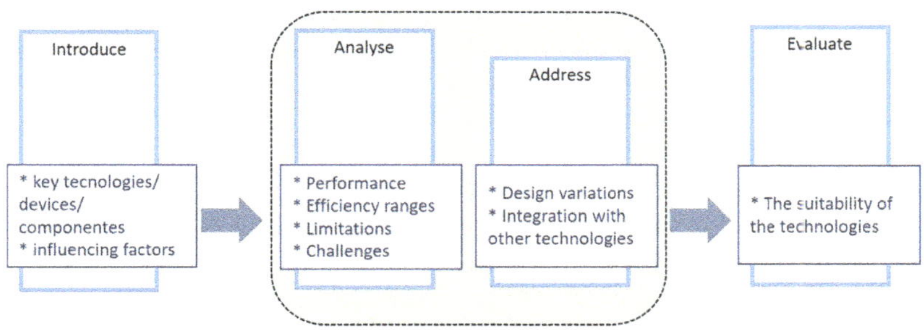


**Figure 12.7.** Flowchart for the assessment of the sustainability of the adopted technologies.

For Cabeza and Chàfer (2020), the ZEB concept was originally based on a conjunction of the ideas of bioclimatic architecture, efficient lighting systems, a preference for renewable energy sources, water consumption management, equipment and facility automation, as well as strict consumption control (figure 12.6).

Cabeza and Chàfer (2020) also argued that when new technologies are implemented in buildings, they should be monitored to guarantee their total efficiency, highlighting that the energy demand must first be reduced in order to subsequently reduce consumption (figure 12.7), and ensuring that the interventions can be technological, behavioral, or regulatory.

Torcellini *et al* (2006) defined a (net) zero-energy building as an energy-efficient building in which electricity is considered to be the energy source, as supplied by the grid, and the amount of renewable energy yearly exported by the building to the power grid must be less than or equal to the amount of energy consumed from the power grid.

The authors also define four ways to achieve the zero-energy state in buildings through low energy consumption, commonly called low-energy buildings, which are:

- *zero site energy*, defined as the potential to generate electric energy for a building using the resources present on site, based on the energy it uses in a year. In this approach, the power consumed by air-conditioning systems, heating, ventilation, equipment loads, and the lighting system is minimally evaluated;

- *zero source energy* considers the entire annual energy production chain as well as the primary energy consumption of the building. This assessment considers electricity, the fuels used in the processing and transportation of materials and components to the building site, among other aspects of the imported and exported energy that are multiplied by the appropriate site-to-source conversion multipliers;
- the *zero-energy cost* definition evaluates the ratio (which should be at least equal) between the total amount of money that is collected from the sale of energy produced on site to the energy provider, and the total amount paid for the use of services and energy consumed throughout the year;
- *zero-energy emissions* checks whether the amount of renewable energy used that is free from greenhouse gas (GHG) emissions is at least equal to the amount of energy obtained from GHG-emitting energy sources.

Marzsal *et al* (2011) exemplified the five renewable supply options related to Nearly Zero-Energy Building (NZEB) definitions, based on the location and the production of renewable energy, as depicted in figure 12.8. The authors associate self-sufficiency with types I and II, also as NZEBs, where renewable energy is available on site, while types III, IV and V refer to buildings that mainly use renewable energy generated outside the place of use.

Kurnitski *et al* (2011), based on the European standard EN 15603:2008, recapped the attributes of the nearly zero-energy concept—the optimized cost, the technical use of primary energy, and the primary energy percentage covered by the energy generated

**Figure 12.8.** The renewable supply options related to NZEB.

from renewable sources. Another very common feature in these buildings is the use of solar thermal and/or photovoltaic energy systems. A recent decline in the costs of implementing solar cells, associated with the rising costs of conventional sources of energy, has contributed to a substantial increase in the use of this technology (Athienitis and O'Brien 2015). According to these authors, these two characteristics highlighted for solar energy are important factors in meeting the definition of the nearly zero-energy concept.

A balanced-energy building presupposes an architecture suitable for using construction elements and equipment of high energy efficiency, combined with the availability of electric energy from renewable sources. The reduction of energy consumption in new buildings can be achieved through projects that integrate energy production technologies with the adoption of energy-efficient solutions and the use of energy-saving programs.

Athienitis and O'Brien (2015) stated that the planning of a zero-energy building should consider the integration of energy strategies and passive solutions; building optimization applied to design, execution, and operation; and the use of solar energy technology as a resource, among other forms of energy production, considering the building typology and solutions that take advantage of the availability of solar energy for urban areas, solar heating, and natural light.

## 12.4 Programs and regulations involved in the change to reduced energy consumption

Feng *et al* (2019) showed that policies encouraging the adoption of energy efficiency ideas have been stipulated in several places, mostly in the United States and European countries. The programs and policies they cited are under development and continuously monitored for effectiveness. Other programs have been interrupted or renewed with new goals, since their previous purposes have not been achieved. The most representative programs are described in table 12.1.

Observing the most developed programs to date, it is possible to verify that the targets set by the countries mentioned in table 12.1 were built over many years. It is clear that these actions are a political priority around the world, as seen by the importance of the energy policies and their revisions mentioned in the later examples. Some of these regulations were decisive in increasing renewable energy use, which has reached a level of approximately 20% on the European continent. However, there will be a need for changes in some policies in order to reach the target set for 2030, which states that 32% of the energy used should be renewable energy.

## 12.5 Zero-energy-building examples

The examples presented in this chapter are buildings located in contrasting regions such as tropical and temperate climates, and are based on studies developed by Cabeza and Chàfer (2020), whose synthesis is represented by figure 12.9.

**Table 12.1.** Energy efficiency programs, policies, and regulations in developed countries.

| Country/ Region | Organization | Program | Current state |
|---|---|---|---|
| Europe | Directive on Energy Performance of Buildings (EPDB) | ZEBRA 2020 (2010) | The results published in 2016 were apparently concluded after confirming the presence of important barriers to the implementation of the objective. |
| Germany | Directive on Energy Performance of Buildings (EPDB) | Act for the Promotion of Renewable Thermal Energy (Released in 2010; updated in 2017) | An act was promoted in favour of using thermal energy as a renewable source. This law is part of a set of laws (EEG) originally proposed to encourage the production of electricity in Germany. The result evaluated today remains uncertain. |
| France | Ministry of Environment, Energy, and the Sea | Act for Energy Transition for Green Growth (2015) | Created after France's COP21 participation, the act was published in order to establish goals for France's growth. In 2016, the targets were revised to include a doubling of the share of renewable sources in the next 15 years (2030), and to promote the integration of renewable sources into the electricity grid. |
| Denmark | The Ministry for Climate, Energy and Buildings | Building class 2020 (2015) | To date, in 2018, the regulation of energy requirements for public buildings has been published. In 2014, the country's energy efficiency action plan was published with targets for reducing primary energy consumption in 2020. |
| United States | U.S. Department of Energy (DOE) | The Building Technologies Program (2008) | This program had as its final objective, in addition to that mentioned by Feng *et al* (2019), the establishment of an activity program plan made for the quadrennium 2008–2012. After the beginning of this program, several construction codes were also instituted, which aimed for energy efficiency. |
| Canada | British Columbia (BC) Energy Step Code Council | BC Energy Step Code (2017) | In the 14th revision, the energy code prerequisites required by the Canadian state for new buildings (from December 2019) was advanced, which already classified uses, climates, and even reference models for thermoenergetic evaluations. |
| United Kingdom | Ministry of Housing, Communities & Local Government | National Planning Policy Framework (2012) | In April 2019, the UK Green Building Council (UKGBC) published a document that included the definitions for UK zero carbon buildings. Some cases (residential and non-residential) were identified by the institution as office buildings with low energy consumption certificates (NZEB). |

**Figure 12.9.** Cycle of implementation strategies for a zero-energy building.

### 12.5.1 Malaysian green technology corporation—a zero-energy office

This building is considered to be pilot project; it includes a research laboratory for the development of buildings with sustainable offices in tropical climates. The zero-energy office (ZEO) building uses active and passive energy concepts and systems integrated with the office building.

It is located approximately 40 km south of downtown Kuala Lumpur and was completed in October, 2007. The climate of the region is classified, according to Köppen and Geiger, as Af (a humid tropical climate); its latitude is 3°08′28″ N, its longitude is 101°41′11″ E, and its height above sea level is 56 m. The orientation allowed the window placement to face towards the north and south in order to reduce solar heat gains. Volumetry with a staggered section was proposed for self-shading.

The ZEO is lit mainly by daylight and is only aided by artificial lighting at night and during darker cloudy periods. The optimization of the building's natural lighting was carried out through simulations. The building's envelope is made from walls and roofs with thermal insulation as well as windows with low-emissivity glazing. In order to prevent solar radiation from entering the interior, high-performance glass and double-sealed glass were adopted.

As the thermal insulation was implemented, another technology was applied, which was a cooling storage system based on an enthalpy wheel, allowing the air-conditioning load to be reduced, since most of the load is created by dehumidifying the air. The cooling is stored overnight in concrete slabs and in a phase-change thermal storage tank. Pipes embedded in the concrete floor slabs release cooling to the rooms above and below, minimizing the use of air conditioning.

Daylight autonomy achieves 95% during working hours. Due to the optimization of the lighting system, which simulates natural light, energy-efficient lighting and lighting control are possible, resulting an installed lighting load of only 5 $W/m^2$.

This natural lighting reduces the power demand enough for the ZEO's power to be supplied by its own Building Integrated Photovoltaic System (BIPV), which is composed of photovoltaic panels connected to the grid. Solar energy powers the grid during the day and is purchased at night to operate the air conditioning. Four different photovoltaic technologies are installed: on the main roof, the photovoltaic technology consists of 47.28 kWp polycrystalline modules; the secondary main roof has 6.08 kWp amorphous silicon modules; the lobby includes semitransparent glass photovoltaic modules rated at 11.64 kWp; and the roof of the parking lot includes 27 kWp monocrystalline photovoltaic modules.

Supporting the photovoltaic system, the roof supports a trickling night cooling mechanism, which works as the chillers run only at night; heat is released from the PV roof by trickling water over the slope of the roof. Water is added at the ridge of the roof and collected at the gutter, releasing heat from the wet room through radiation to the environment through convection and evaporation. This process can release heat at relatively low water temperatures ranging from 25 °C to 35 °C.

The cooling load of the building is very low compared to that of a traditional building; thus, the amount of heat released is equally small. This low load state allows the use of the trickling system, since a building with a higher load might not have the conditions required to use the roof as a device to cool down the water.

The facility also incorporates sustainable design features, such as the use of rainwater to meet two-thirds of the water needs. Other active measures have been implemented to reduce the need for electricity, such as high-efficiency pumps, fans, and office equipment.

Since October 2007, when the Malaysian Energy Center moved into the building, the building's performance data and its energy systems have been monitored continuously. The building's design focused on energy efficiency (100% daytime lighting, slab cooling, double glazing, among others). The building is self-sufficient due to the use of energy from the BIPV systems and achieves a desirable energy index of 35–40 kWh $m^{-2}$ $year^{-1}$, considering a useful area of 4200 $m^2$, occupied by 111 employees.

### 12.5.2 BedZED (Beddington zero fossil energy development)

BedZED is a mixed-development urban brownfield in Hackbridge, in the London Borough of Sutton. The Borough of Sutton provides 82 rooms in a mix of apartments, maisonettes, and semi-detached houses and approximately 2500 $m^2$ of office space and community accommodation, including a health center, nursery, coffee shop, and sports club. The community was designed to host around 240 residents and 200 workers. BedZED is Britain's first carbon-neutral urban development.

It is located approximately 17 km south of London and was completed in 2002. The climate of the region is classified, according to Köppen and Geiger, as Cfb (an oceanic climate) and it is at a latitude of 51°22′55.5″ N, a longitude of 00°09′21.67″W, and 27 m above sea level.

In order to minimize energy demand, the houses face south to make the most of the heat from the Sun, and are equipped with high levels of insulation, windows with triple glazing, and the latest energy-saving appliances, including energy-saving water devices. This solar orientation allowed the envelope to combine high insulation, heat recovery, and a passive solar gain stored inside each apartment by thermal floors, walls, and internal gains. A comfortable temperature is achieved most of the time, without any need for backup heating. In summer, the work spaces placed in the shaded areas of the housing terraces facing south allow all apartments to have outdoor garden areas with good access to sunlight, while at the same time allowing for cross ventilation.

The lighting design of BedZED aimed to reduce the electrical demand; the apartments were fitted with energy-efficient appliances:

- a 20-watt compact fluorescent light bulb;
- A-rated fridge/freezer and washing machine;
- visible meters to make residents more aware of consumption;
- good daylight design to reduce the need for electric lighting.
- passive ventilation to remove the need for electric ventilation or fans;
- aerated showers to avoid the need for power showers.

Around 20% of BedZED's electrical demand is met by on-site photovoltaics. They were originally installed to power electric cars, but since this market has not yet taken off, this energy is being used in the buildings. The system is composed of 777 $m^2$ of photovoltaic panels, making up 1138 laminates on the rooftops and in the south-facing second-floor windows. It was intended to power 40 electric vehicles during the ten-year target period; however, the uptake of electric vehicles was much slower than that hoped for. Currently, the electricity they produce is used by the site as a whole, with any surplus supplied to the grid.

The total size of the array is 108 kWp; it was designed to provide 88 000 kWh of electricity annually, accounting for 30% of the whole site's consumption; the $CO_2$ reduction includes a 20% contribution from the photovoltaics. Also, the supply of hot water is provided by district heating; the hot water system contributes to better energy performance. The building is considered to reduce its ecological footprint by two thirds, including a 90% reduction in carbon emissions. BedZED's renewable energy was to be generated by a combined heat and power (CHP) plant using locally sourced woodchip to generate heat for both hot water and electricity. The plant was operational for long periods when switched to a gas supply. However, it was switched from gas back to a biomass pellet boiler in 2017, which provides heat for the district heating network sourced from Spain, not from local wood waste.

One of the upsides of the BedZED households is that they use 2579 kWh of electricity per year, which is 45% lower than the average in Sutton. When the biomass CHP plant is not in use, BedZED uses gas to power the district heating system. On average, households use 3526 kWh of heat (from gas) per year—81% less than the average in Sutton. It was found that the residents only use 72 litres of mains water per day, topped up by 15 litres of recycled or rainwater. This is less than half of the local average.

The boiler consumption ranges from 277 to 9424 kWh. In 2008, the BedZED average of 3526 kWh $dwelling^{-1}$ $year^{-1}$ was 81% lower than the Sutton average of 18 924 kWh. However, this figure was not based on meter readings but on the estimated total site consumption by the mains gas boiler and woodchip CHP which was apportioned down to a three-bed dwelling. Waste heat from the engine jacket and exhaust is tapped by heat exchangers and aims to provide all the hot water and heating needs for BedZED. BedZED's design reduces space heating by 88% (2003 monitoring), therefore the CHP mainly supplies hot water for washing, for which it meets the daily total demand.

## 12.6 Conclusions

This chapter addresses issues inherent to energy consumption and construction technologies that promote efficient buildings. Its primary goal is also to clarify how government actions and programs solve problems related to this topic, and how society as well as the civil construction industry meet the program requirements.

Compared to European countries, the greatest difficulties are observed in developing countries, especially those located in the tropics, with regard to the development of policies that promote the adoption of measures to reduce energy consumption related to buildings. Even though this is a changing scenario, countries in the tropics that adopt such measures remain few.

Aside from government efforts, being energy efficient also means meeting standards and regulations with the goal of setting higher performance requirements for buildings, and assigning credits to each level reached according to the efficiency achieved after evaluation. Problems mainly linked to climate change drive the energy situation to an even more delicate point, since the outcomes of implementations in response to these changes require time not only to be noticed, but to take effect, in addition to the time required for policies to be created and applied.

Researchers have pointed out measures for increasing energy efficiency, methods for reducing primary energy consumption, and the implementation of several sources of energy, emphasizing that the paths for energy production must change towards sustainable solutions and the adoption of other renewable energy sources.

Examples of buildings across the world, particularly those seen as references, already corroborate the effectiveness of passive and active methods against the increase in energy consumption. Traditionally, buildings which use these methods associated with the zero-energy concept are frequently observed in temperate climate environments and, more recently, in tropical climates. This is relevant when we consider that the measures adopted in these buildings can be reused and be adapted to contexts where there is a greater need to reduce consumption and increase efficiency in terms of energy usage.

Finally, it is also noticeable that this process occurs through dialogues with society and investment in research and development in order to make measures that reduce the impact of energy consumption in the built environment understandable and concrete. The policies created can be considered ineffective without the vital collaboration of people and industry, which are responsible for a large part of the energy consumption in buildings today. Technological advances in materials are of paramount importance, which demonstrates the importance of funding science and research, which is considered a fundamental action to reduce the impact of buildings on the environment in the next few decades.

## Bibliography

Athienitis A and O'Brien W 2015 Net ZEB case studies. *Modeling, design, and optimization of net-zero energy buildings* (Hoboken, NJ: John Wiley & Sons, Ltd) pp 241–350

Brunetta R F D S and Souza S N M D *et al* 2020 Life cycle energy assessment and carbon dioxide emissions of wall systems for rural houses *Ambiente Construído* **21** 37–50

Cabeza L F and Chàfer M 2020 Technological options and strategies towards zero energy buildings contributing to climate change mitigation: a systematic review *Energy Build.* **219** 110009

Dascalaki E G and Argiropoulou P *et al* 2021 Analysis of the embodied energy of construction materials in the life cycle assessment of Hellenic residential buildings *Energy Build.* **232** 110651

Feng W and Wang R *et al* 2019 A review of net zero energy buildings in hot and humid climates: experience learned from 34 case study buildings *Renew. Sustain. Energy Rev.* **114** 109303

Kurnitski J and Saari A *et al* 2011 Cost optimal and nearly zero (nZEB) energy performance calculations for residential buildings with REHVA definition for nZEB national implementation *Energy Build.* **43** 3279–88

Marszal A J and Heiselberg P *et al* 2011 Zero Energy Building - A review of definitions and calculation methodologies *Energy Build.* **43** 971–9

Monzón-Chavarrías M and Guillén-Lambea S *et al* 2021 Heating energy consumption and environmental implications due to the change in daily habits in residential buildings derived from COVID-19 crisis: the Case of Barcelona, Spain *Sustainability* **13** 918

Ramesh T and Prakash R *et al* 2010 Life cycle energy analysis of buildings: an overview *Energy Build.* **42** 1592–600

Torcellini P and Pless S *et al* 2006 *Zero Energy Buildings: A Critical Look at the Definition* (Golden, CO, US: National Renewable Energy Lab NREL)

Weißenberger M and Jensch W *et al* 2014 The convergence of life cycle assessment and nearly zero-energy buildings: the case of Germany *Energy Build.* **76** 551–7

**IOP** Publishing

Sustainable Urban Development
Topics, trends and solutions
**Luís Bragança, Cristina Engel de Alvarez and Luisa F Cabeza**

# Chapter 13

# Energy use and the implementation of renewable energy in cities

**Norberto S Odobez, Carlos Godoy and Luisa F Cabeza**

This chapter addresses the implementation of renewable energy sources (RESs) in cities, describing, for example, those that could be applied in the different spaces of the city, such as residential buildings, single-family homes, office and commercial buildings, small and medium-sized businesses, public buildings, and in suburban areas or spaces outside the city where there are normally industries, small condominiums, or isolated urban areas that can be viewed as islands of consumption and/or generation from a certain perspective.

RESs are analysed, evaluated, and compared, together with a discussion of the way in which they are used to contribute to the sustainability of cities; this chapter also considers the implementation of renewable energies through the self-generation of energy for self-sufficiency and distributed generation, both for the entire city or sectors of the city, based on the various available technologies and experiences in different cities.

## 13.1 Introduction

Energy use and consumption in cities is not isolated from other phenomena or activities that occur in them, but they are interdependent and dynamic. Moreover, from an energy point of view, there is no way to equate cities and contemplate or evaluate problems in the same way.

Cities are 'living energy entities'; therefore, we address how we can intervene in the energy consumed by the city. This should be done by a better design that minimises energy demand (also known as sufficiency), with increasingly smaller consumption, rational use, and energy efficiency (EE) and finally, supplied by RESs.

These approaches are linked to the geographical location of the city and the availability of RESs, considering their dependence on other resources. Therefore, cities can leave the context of their province, region, or country to generate their own energy, or encourage the use of RESs. The possibility of applying policies (local or national) for the use and generation of renewable energy depends on the economic resources available in a city, the idiosyncrasies of its inhabitants, and, of course, the size of the city in relation to, for example, the density of its inhabitants (inhabitants per square kilometre) (Urge-Vorsatz *et al* 2020).

According to the International Energy Agency (IEA), the building and construction sectors combined are responsible for more than a third of global final energy consumption and almost 40% of total direct and indirect $CO_2$ emissions. The energy demand associated with buildings and building construction continues to rise, driven by better access to energy in developing countries, increased ownership, the use of energy-consuming devices, as well as rapid growth in the acreage of buildings worldwide.

To define the potential types of RES that can be applied, they can be divided between those that are variable sources (solar photovoltaic, solar thermal, and wind) and those that are continuous (biomass, biofuels). Energy increasingly needs better management that can integrate both demand and availability in a holistic way (including storage, which is becoming a key technology).

Taking this situation into account, actions that go beyond the basic technologies used for generation come into play. Other disciplines then appear that change the way in which energy is managed and used, such as grid automation, proactive users, microgrids, and the interconnection of electrical and non-electrical networks (for example, the use of existing networks with new energy vectors, such as hydrogen, to achieve the final concept of smart grids).

Resilience thinking is important for achieving sustainable communities. If sustainability involves maintaining a desired state of generation and supply, in other words, the transmission and distribution of energy required for the effective functioning of urban systems, resilience to disasters will be a prerequisite (Sharifia and Yamagata 2015).

Therefore, renewable energies are an effective alternative tool for achieving this resilience, due to the possibility of dividing, differentiating, and becoming independent from a general power supply network or interacting with it in a safe and harmonious way, given the new technological instruments and equipment that are being developed to link them, such as using a 'reliable and open energy services platform' (Intedya 2017).

## 13.2 Short general description of renewable energy sources that can be used in cities

### 13.2.1 Description of renewable energy sources

#### *13.2.1.1 Solar energy*

Within RESs, solar energy is the most abundant and easiest resource to adapt to the urban environment. The use of solar roofs is a very effective way to take advantage of this resource, without requiring urban land costs for its installation. Its effectiveness and impact on the energy system depend on the climate and latitude.

*Solar photovoltaic (PV)* is the easiest to use, although it is limited by its cost, characteristics, and regulations. The solar panel format has proved to be flexible enough to adapt to existing buildings. The incorporation of power-generating cells into the windows of office buildings is an ongoing technological process. The initial investment cost is an important barrier, especially in areas where the climate is not completely favourable to exploitation, such as cities near the poles, or cities with a temperate climate, where it also produces a completely unacceptable generation characteristic curve for a network at the intended scale of integration. Like wind generation, it needs to be effectively implemented to form a hybrid generation system supervised by a new concept in load management, which comes hand in hand with the modernisation and updating of electrical distribution systems, which, to a great extent, continue to be oriented to the characteristics of high inertia and passive consumers. Solar PV system costs have continued to fall in recent years, making this technology more attractive for use.

*Solar thermal energy* is an important heat source for district heat and electricity generation (SE 2019). It is categorised into two key collector technologies, concentrated and non-concentrated. Solar concentrator systems have better efficiency and greater heat production. However, their installation and use for electricity generation is not effective in an urban environment, since it is difficult to place them due to the large collection area needed; they are more efficient in cheaper land areas, such as those at the administrative limits of cities or directly in desert areas. Another factor to take into account for solar concentration collectors in the urban environment is the design of the solar park and the working methods, given the importance of controlling the effects of loss of solar reflection and the requirement not to disturb the safety of other activities during the day, such as air navigation (Shilston and Danks 2018).

In contrast, the use of non-concentrated technologies such as flat plate collectors or vacuum tube systems as heat sources for domestic hot water (DWH) and space heating, or cold for air conditioning systems (by absorption), especially in hot climates, has better results. These systems provide DHW to individual low-power installations at up to 120 °C in pressurised systems, which are compact and easy to install on the roofs and facades of houses and buildings. They are also scalable for use in commercial buildings, allowing the application of district heating networks at medium temperatures between 60 °C and 70 °C (SE 2019).

#### *13.2.1.2 Wind energy*

Among the challenges facing the implementation of wind energy use are the visual impact of the generators, the lack of sustained winds combined with disturbances to the flow of air caused by buildings, and the lack of space to implement the wind turbines. Therefore, the design of new generators seeks a compromise between their visual appearance, architectural integration, and operational flexibility (Barragan-Escandon *et al* 2019). For example, among the available technologies, the iconic horizontal axis generator, which is the design with the best conversion efficiency, suffers a considerable drop in performance in the urban environment due to constant changes of wind direction and intermittent winds (Keirstead and Shah 2013).

To overcome this problem, the use of vertical axis generators takes advantage of the winds produced by the turbulence caused by particularly tall buildings; in addition, their profile allows a simpler architectural integration at the cost of lower performance. Nevertheless, without the availability of permanent winds in the installation area, such generators can hardly be implemented as a single solution without complementing them with another energy system to form a hybrid system (www.kliux.com).

#### *13.2.1.3 Hydraulic energy*

The use of hydraulic resources depends on the geography of the urban location, which, in general, is characterised by flat areas. However, in mountainous areas or those with large rivers, microgenerators (microturbines) can be used to produce energy in conjunction with energy reservoirs (for energy storage). A particular use case would be that of a reversible microturbine with a special design to resist entrained wastes, which is used as a control element in underground storm water reservoirs, where the energy available from a controlled discharge of water would be used when river levels are low, or as a bilge pump in the event of temporary increases in water level due to meteorological phenomena.

Another type of hydraulic energy is tidal wave energy, which requires proximity of the city to the sea. In general, tidal projects require a very large scale to be implemented efficiently, which involves nation-level projects that are beyond the budget of a coastal city. In particular, this category includes projects based on the enclosure of water at the change of tides, those that take advantage of the turbulence caused by waves, and those that use underwater currents (Keirstead and Shah 2013).

#### *13.2.1.4 Geothermal energy*

Geothermal resources are classified into high- and low-temperature systems. The implementation of high-temperature systems, which are the most interesting to work with, requires the availability in the immediate vicinity of a resource that, due to its unstable nature, is opposed to the conditions for establishing a city (that is, it is the logic of establishing a human settlement at the foot of an active volcano), which explains the small number of places that meet the conditions required to take advantage of this resource. Low-temperature systems are suitable for the provision of space heating, DHW, and air conditioning in buildings when connected to a reversible heat pump.

#### *13.2.1.5 Biomass*

Biomass is a term used to describe any substance that comes from animal or vegetable matter that is used as fuel, including wood, edible crops, clippings of plants and grass, forest residues, and the organic components of urban and industrial waste. All organic materials that have been geologically transformed into fuels such as coal, oil, and natural gas are excluded (NETL 2020).

Biomass is generally obtained as a solid and it is difficult to use in any application without substantial physicochemical modification. The direct use of biomass occurs when it is dried and compacted under pressure, forming pellets or briquettes for use as a substitute for firewood or charcoal in the production of heat for the conditioning of homes or kitchens, provided that the burners used have high efficiency.

The conversion of solid biomass into liquid and gaseous fuels has significant advantages in handling and final application. A wide range of processes (including biological processes that produce ethanol or methane and thermal processes for heat production) is available to convert waste and biomass into solid, liquid, and gaseous fuels, from which a great variety of byproducts can be obtained, including electricity (Bridgwater 1995).

The thermo-conversion of biomass include all the techniques in which the transformation of biomass occurs at high temperatures. A gasifier transforms organic matter via pyrolysis (gasifying biomass in the absence of oxygen) using heat and pressure, decomposing it into solids (such as ash and salts) and combustible gases (such as hydrogen, carbon monoxide, and methane). This gas, called Syngas or synthesis gas, can be used to produce other products or can be treated for combustion in a conventional system. Its main disadvantages are its cost and the impossibility of using this gas in natural gas networks. In a conventional system, the derived gas is burned immediately and is used to produce steam for an electric generator, which is cleaner and more efficient than if the biomass were directly burned.

The main advantage of thermo-conversion, compared to the biological method, is the rate of biogas production. Therefore, a derivative is the coupling of both processes for an even more efficient conversion of biomass using thermal conversion to treat waste from bioconversion plants, extracting energy stored from lignin structures. It should be noted that modern gasification technologies originate from the exploitation of coal, gas, and oil sands, so they are mature techniques in the industry (NETL 2020).

In bioconversion, bacteria break down the sugars in the fibres of organic waste into alcohols and gases such as methane. This process occurs in a controlled manner in biorefineries, in containers capable of storing fuel elements and mixtures for long periods. The contents are then separated and filtered. The gas obtained can be introduced into the natural gas networks and the refined alcohols can be used in pure thermal machines or in a mixture with conventional fuels. It is also produced in landfills, where the gas is collected through previously planned and installed pipes, to be used in conventional generators specially adapted to this fuel. The great difficulties of these systems are the correct classification of biodegradable materials from the mass of urban waste, something that is solved in advanced countries but difficult to apply in developing countries; the considerable odour that they emit during the storage and preparation of the mixtures; and the subsequent disposal of the biogases produced, making their installation near populated areas problematic.

#### *13.2.1.6 Municipal solid waste and landfill gas*

The use of municipal solid waste (MSW) for energy production is a good way for municipalities to take advantage of this resource. Its use can be carried out by direct burning, and from there, the generation of thermal and/or electrical energy; alternatively, it can be carried out by the biodigestion of plant waste by means of biodigesters for the production of biogas. In particular, if the alternative is to landfill MSW, landfill gas (LFG) production is the technological approach most often applied to obtain energy. These two alternatives for using waste resources to produce energy, although not considered to be renewable energies, could be used as an energy alternative in a municipality or group of municipalities to reduce the amount of waste that is landfilled, using it for other purposes and also making a contribution to the environment.

### 13.2.2 Application of energy efficiency and renewable energy sources in cities

The application of renewable energy in cities depends on the resource that the city has; therefore, the available sources in the city's location should first be identified.

The methodology that should be followed is presented in figure 13.1. The city should consider the proximity of each resource in the context of the deployment of RESs. These resources should be quantified in the best possible way to ensure that the decision regarding the application of these technologies will be the most efficient and appropriate.

Policies and legislation should also be mentioned in the case of deployment of RES in cities, considering 'The Right to the Sun', and taking into account that most of these projects are long-term and their economic viability needs to be considered. Moreover, obtaining the required permits may take a long period of time, and in this period, new buildings or structures may appear, jeopardising the project itself. Cities are dynamic, and their fast or slow changes have a big impact in any project seeking to implement RESs in cities.

First, **the city's own resources** should be considered, among which, solar, wind, and local resources (i.e. wastes) should be considered first. Small-scale solar PV could be

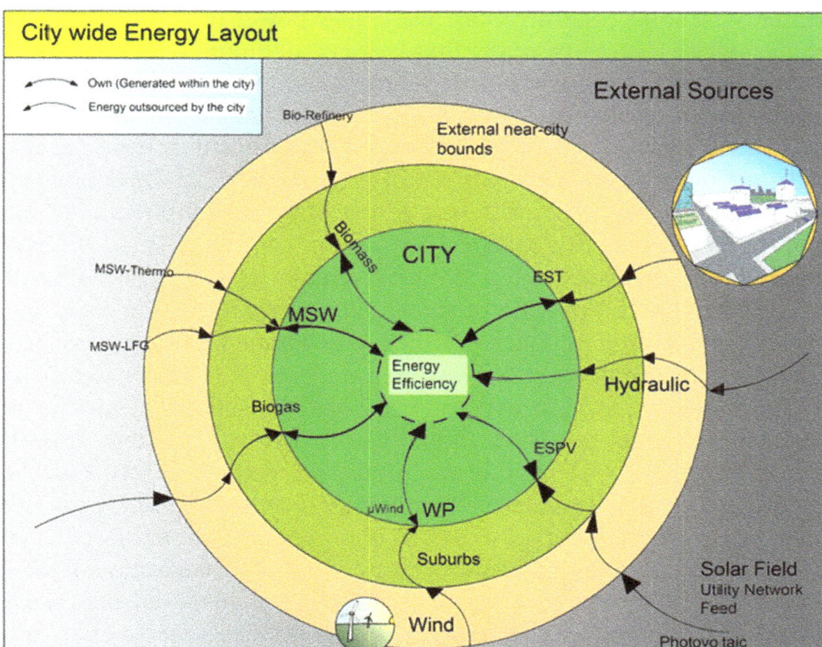


**Figure 13.1.** Available resources in cities using RESs. Source: designed by the authors.

applied here. Today, special panels are available for 'solar roofs', which offer high performance and efficiency in relation to the occupied surface, for example, for self-supply, direct injection into the grid, or as part of an urban microgrid. Similarly, but perhaps to a lesser extent, for the reasons stated above, wind technology could be considered.

The use of solar thermal technology on roofs, which offers various thermal capacities, is a good possibility. This can use different technologies, such as flat or vacuum tube collectors, which are better adapted to various regions and climate conditions. This is used mainly for DHW, although space heating is also possible. Local products, such as waste and green waste, which can be used as biomass or for the production of biogas, are a possibility; their treatment could be performed in areas close to the city for the reuse of the energy in the city itself.

**Nearby resources**, which can be obtained from the vicinity, would be those obtained in facilities or properties on the periphery, where there may be industries and free space. Here, energy can be produced on the roofs of factories, using PV connected to the grid or connected to a city microgrid. Other sources are medium-power wind or the transformation of industrial sites' own wastes, which can be used as a complement to forest or organic biomass. Finally, the use of nearby non-productive freeholds can combine energy and food production.

**Nearby resources**, at a distance that leaves the periurban area, i.e. smaller cities about 15 km from the city centre. Here, small or medium-power hydraulic energy could be installed, depending on the geographical location of the city. Also, wind and solar farms as well as biomass and high-power waste could be used for electricity generation. Resources could be used for district heating, for example, from solar thermal energy by solar concentration. These new power plants could also interact at the public service or 'utility' scale, that is, generally more than 5 MW.

Here, the concept of the **city's energy core** is used for the adoption of energy-saving technologies or the rational use of energy. This should be applied at the public or private level, as a preparatory step in the transition from fossil fuels to RESs and to comply with challenges such as the Sustainable Development Goals (SDGs) by 2030; in this case, SDG 7, which aims to double energy efficiency worldwide (https://sdgs.un.org/goals).

Although this seems obvious, it is not what usually happens. The efficient use of energy should already be common practice, but this is not yet the case; the reason may be non-compliance with policies or the use of erratic ones. As published by the World Bank, 'Of all alternatives that exist to satisfy the demand for electricity in the future, almost always the least expensive are the measures that aim to improve energy efficiency'. Energy efficiency should be the core of the energy transition.

We can more broadly incorporate the concept of energy performance, which is more encompassing than the simple vision of energy efficiency. Energy performance improvement refers to the set of possible improvements in terms of energy use (optimisation of the way energy is used), energy consumption (reduction of the amount of energy consumed), and energy efficiency (technological improvements) (GIZ 2018).

On the supply side, and specifically speaking of utilities, work would be done to intervene in inefficiencies and energy losses; national standards should be developed for the energy efficiency of resources (energy-saving goals for utility companies).

In the city itself and its constituent elements, such as infrastructure, users, and consumers, the intervention points could be energy codes for new buildings, standards for household appliances, programs to promote comprehensive modernisations of the energy systems of constructed buildings, and energy efficiency actions, such as building renovation and building envelope improvement and a reduction of heat losses by the use of insulation or the reduction of thermal bridges.

Other technical solutions include high-performance building roofing; efficient heating and cooling; solar control, shading, and passive solar heating; efficient lighting and increased daytime lighting; and smart system control technologies that optimise energy use within the building.

Municipalities should emphasise integrated urban planning and urban designs to maximise energy efficiency and improve the quality of life where people work and live. Examples include increasing building density in urban areas, developing transit corridors, and incorporating mixed-use spaces. Infrastructure planning should be integrated, in order to ensure that the energy cost and efficiency implications of other systems (i.e. roads, water, heating, and cooling) are adequately considered (Singh 2016).

### 13.2.3 Other technologies that contribute to the application of renewable energy sources (RES) in cities

#### *13.2.3.1 Advanced management*

To obtain a sustainable and quality energy supply, new management systems are required, to address the intermittent nature of solar and wind sources. This implies the development of a new concept of grids by changing the relationship between generators and consumers. Energy storage and load management are included within the energy management system.

#### *13.2.3.2 Energy storage*

The key requirements for energy storage that are applicable to a city are energy density, resistance to loading and unloading cycles, power, and application security. Modern energy storage schemes are categorised, depending on the operating characteristics, as those for short (on the order of seconds), medium (minutes), and long (hours) periods. Examples of some available technologies are the promising flywheel technologies for instantaneous frequency regulation and melt-state electrode batteries.

A flywheel is a mechanical energy storage device that can be used instantly. It essentially consists of a rotating mass driven by a reversible motor system; when necessary, the rotating mass delivers electrical energy in a similar way to that of a turbine and recharges itself by taking energy to increase its speed again. This technology is very beneficial for improving network stability, since it captures energy from intermittent sources and returns it at critical moments, thanks to the possibility of responding instantly to stabilise the frequency (ESA 2020).

Sodium sulphur (NaS) melt-state electrode batteries were developed in the mid-1960s and operate at high temperatures ranging from 300 °C to 350 °C. The active materials in these batteries are molten sulphur (the positive electrode) and molten sodium (the negative electrode). These are separated by a sodium aluminate ceramic that works as an electrolyte, allowing only positive sodium ions to pass through. During the discharge cycle, the sodium delivers an electron to the negative terminal, passing through the solid electrolyte to the sulphur compartment, which receives an electron through the positive terminal, forming a polysulfide. During the discharge cycle, the process is reversed. Its main drawbacks are the high melting temperatures of the electrodes, due to technical and safety difficulties, since they require their own independent heating system. On the other hand, these batteries are highly efficient, have a high energy density, have a very high number of charge cycles, and are a proven technology in frequency control systems and peak demand trimming (https://www.arup.com).

Ni-Cd batteries are renowned for their durability, ease of application, and long-term reliability, which offset their intermediate energy density and cost. Available since 1910, they were kept alive by improvements in electrode manufacturing. In this type of battery, nickel is used for the positive electrode and cadmium is used for negative electrode. They are used in renewable energy facilities as energy reservoirs due to their loading and unloading capacity, the high level of knowledge about this technology, and their safety (ESA 2020).

Lithium-ion batteries emerged from their use in portable consumer electronics. They are characterised by their durability, flexibility, energy density, and portability. The development of new power cells seeks to feed new applications, such as their use in mobility and storage to stabilise renewable energy grids, while reducing manufacturing costs. The name 'lithium-ion' refers to a whole range of electrodes which are not made of pure lithium, but chemical compositions of lithium and lithium phosphate in the positive electrode and graphite or lithium titanate in the negative electrode, among others. This flexibility provided by the technology gives lithium technological centrality, which ranges from applications of a few kilowatt hours for solar roofs to several megawatt hours for the stability of electrical distribution networks.

Pumping stations are the most widespread solution in the industry. These pump water from a lower elevation for storage in a higher reservoir. At periods of high energy demand, energy is delivered by releasing the water from the reservoir through the turbines, as in a conventional hydroelectric plant. When demand is low, such as at night, at weekends, or at very low-cost times (such as times of excess wind production), specially designed reversible turbines pump the water back to the upper reservoir. The special characteristics of these turbines make it possible to compensate for the inefficiency of the pumping, delivering an overall efficiency of more than 80%; they are very beneficial for the stability of the system and offer economically viable operation due to the peaks and valleys of power pricing.

#### *13.2.3.3 Hydrogen*

Clean hydrogen is obtained from electrolysis, and it is stored to generate electricity again. Today, the efficiency of this cycle is lower than that obtained from other storage technologies. Despite this, the interest in developing its use is due to its energy storage capacity (which is comparable with batteries on a small scale and with pumping stations on a large scale) and the ease of adapting to the technology currently in use (combustion). For sustainable production, alkaline electrolysis is a mature technology that can be implemented in large-scale systems, while proton exchange membrane (PEM) technology is more flexible for implementation in decentralised units. The average conversion efficiency for both technologies ranges from 65% to 70%. The development of new high-temperature electrolysis equipment promises an alternative to PEM and alkaline electrolysis, with an efficiency superior to 90%. The hydrogen produced can be stored in pressurised tanks for smaller energy demands, on the order of a few MWh. For higher requirements, it can be accumulated in a solid state as metal hydride in high-density nanotubes. Large quantities of hydrogen can be stored underground in caverns of the order of 500,000 cubic meters at 190 bars, providing an energy potential close to 100 GWh and even offering the possibility of compensating for seasonal changes in solar-wind production. Electric energy recovery can be carried out using fuel cells at efficiencies close to 50% or in a conventional combined cycle thermal unit, reaching an efficiency of up to 60% (ESA 2020).

#### *13.2.3.4 Thermal energy storage*

The thermal energy captured from the environment or from industrial processes and stored can be used for the comfort of the inhabitants of a building for the period of a day, or even to provide the conditioning needs of an entire season (Cabeza *et al* 2021).

In general, we can say that the most important applications of thermal energy storage (TES) systems are in the construction sector (e.g. DHW, heating, and air conditioning) and in the industrial sector (e.g. heating and cooling processes). For specific requirements, there are mature technologies that are already used in buildings, and others under development with great potential.

TES systems can be installed as centralised plants or distributed devices. Centralised plants are designed to store waste heat from large industrial processes, conventional power plants, combined heat and power plants, and renewable energy plants, such as concentrating solar power (CSP). Their power capacities typically range from hundreds of kW to several MW. Distributed devices are usually intermediate storage systems meant to accumulate solar heat to be used in domestic and commercial buildings (e.g., hot water, heating, and appliances). Distributed systems mainly offer capacities in the range of a few to tens of kW (IRENA 2020).

TES can provide both supply-side and demand-side flexibility and the integration of on-site renewable energy generation.

Future energy systems will need a combination of storage alternatives, as each has a different value proposition in terms of storage period, energy density, and market service, with different response times and operating temperatures, among others. At the city level, stationary storage is used for decentralised rooftop photovoltaic power plants, while at the utility scale, storage is used for electromobility and heat.

Currently, technologies are first classified as active or passive and then according to their energy storage strategy: sensible, latent, thermochemical, or thermo-mechanical heat.

### *13.2.3.5 Description of TES technologies*

Sensible TES systems are the most common, the simplest, and generally cheapest to implement. The energy stored in the structure is proportional to its temperature. The disadvantages are its large size, the necessary insulation characteristics, and the need to use energy to maintain a desired fixed temperature.

A water storage tank is a sensible TES system that stores water at temperature, heated by a purpose-built heater or a solar system to a temperature in the range between 55 °C and 60 °C for intra-daily consumption and at 95 °C for storage. It is especially suitable for domestic hot water installations and scalable up to large commercial buildings, but limited by the available area and the type of container used.

The use of solid materials for sensible TES is a passive configuration in which parts of the structure are arranged by design to store ambient heat in a building. As part of an active configuration, it involves the use of particulate material to store heat from a process such as solar or wind energy.

Underground thermal energy storage is the use of natural or artificial underground geological structures to store heat in an aquifer or groundwater (ATES), in a pit structure arranged to store energy (PTES), or directly in a ground borehole (BTES). The advantages of this system are the large amount of energy it can store and its ability to store both heat and cold between seasons. Its great disadvantage is its low efficiency, which is why it is used in places where an inexpensive source of thermal energy exists (geothermal heat, district heat, snow cold, etc.).

Another technology is the use of latent energy. In this case, energy is stored by changing the phase of matter, so the property of interest is the melting enthalpy. The energy stored is not proportional to the material's temperature, so this implies the advantage of not needing high temperatures for high energy densities, but requires a selection of the correct phase change material (PCM) for the desired temperature. According to Sarbu and Sebarchievici (2018), most of the literature focuses on practical and commercialised PCM materials such as paraffin, recommending and focussing on specialty PCMs with a wide temperature range, such as salt hydrates, and synthesising PCMs that are especially suitable for specific applications.

Ice is an excellent PCM, due to its high heat of fusion, its high heat capacity in its liquid phase, and its non-corrosive (pure) behaviour. This system is relatively inexpensive and is a technology that is already in use. It is configured to use peak energy from the grid or from renewable sources. The system consists of coolers or ice generators, an ice tank and a direct (ice slurry) or indirect (glycol) exchanger system.

The third TES technology is known as thermochemical TES, which includes chemical reactions and sorption systems. These systems achieve considerably higher energy densities than sensible and latent TES, thus allowing designers to consider seasonal energy storage at lower temperatures.

In an adsorption system, energy is stored by breaking the interaction forces between molecules, which is the reason for the low working temperature of this system. It has the advantages of being easily integrated into systems based on thermal tanks, reversible to provide heat or cold, and easy to apply in buildings. Commonly used refrigerants are calcium chloride ($CaCl_2$), lithium chloride (LiCl), lithium bromide (LiBr), sodium hydroxide (NaOH), potassium hydroxide (KOH), and ammonia ($NH_3$). In this type of system, heat from a solar source or recovered waste heat forces the separation and evaporation of water from a solution, producing a more concentrated refrigerant. This is stored, and then re-mixed with water, releasing the stored energy.

Other technologies that cannot be integrated into buildings that could be included inside city boundaries are as follows. For the storage of both electrical and thermal energy, which requires a system of high technological integration and complexity, the use of air compression (A-CAES) and air liquefaction (LAES) is proposed. An air compression system comprises air compressors which are assisted by a TES system based on sensible heat (stone bed, ceramics) or latent PCM to store the thermal energy of the compression process. The air is deposited in deep caverns to economise gas storage under pressure. When energy is required, an expansion process is carried out using a contribution from stored thermal energy, reducing the energy losses of the process.

To solve the problems of the handling and storage of compressed air and improve the overall efficiency of the cycle, the liquefaction process is included in a liquid-air system (LAS), since the transport and storage of liquid air is simpler and cheaper. It also adds a cold storage system, to assist the liquefaction process.

As we can see, the different types of system that can be used for thermal storage mean that, depending on the region or area the city is in, it can adopt the most convenient system in terms of both technology and power scale.

#### *13.2.3.6 Charge-discharge management*

To manage generation peaks and valleys, it is necessary to implement coordination between generators and energy storage systems to adapt the to the user consumption scheme; moreover, consumption automation and electricity tariffs contribute to RES deployment. The implementation of microgrids allows an improved capacity to respond to failures of the main distribution grid, which, in the hypothesis of the self-sufficient city, will always be the backup with low power availability for efficiency reasons. Here, the coordination of generation grids with consumers comes into play through smart grids that manage generation and stability resources. The use of artificial intelligence to control long-term system variables through 'training' that uses possible scenarios based on real data and in real time is a good tool with which to further stabilise the grid against unforeseen scenarios (Tarragona *et al* 2021).

The latest trends in electricity generation point to a new concept of synchronisation and stability using virtual inertia inverters. This implies a revolutionary change, in which low-inertia generators, such as photovoltaic and wind power, can participate in frequency stability. The inverters currently on the market are grid trackers, built on the idea of a frequency network controlled by the inertia of large rotating generators. As more and more renewable power is integrated into the grid, a new synchronisation system is necessary. A solution under study is to change the grid reference using GPS satellites, which would allow RES systems to start by themselves after a blackout.

The possibilities offered by new energy generation technologies or improvements in the costs of those already proven for self-generation, connection to the grid, microgrids, district energy or distributed generation, their management and their application at different scales mean that new actors in society must be defined, as stated in the European Directive EU 2018/2001 in its Article 21 'Self-consumers of renewable energies', its Article 22 'Renewable energy communities', Article 23 'Integration of renewable energies in the heating and cooling sector', and Article 24 'District heating and cooling', among others, to give them the precise characteristics for the application of these technologies.

The use of shared infrastructure between service provider companies and municipalities must be planned, managed, and compensated fairly. However, when possible and appropriate, distributed electricity generation (i.e., power generation at or near the point of consumption) and microgrids will positively influence supply reliability in the event of major storms, floods, and fires, among others. The benefits of decentralised elements for the energy system can be combined with energy efficiency measures on the demand side and with the development of a 'smart' electricity grid (Salmeri *et al* 2017).

## 13.3 Discussion

As described, there are different TES technologies that can be used in cities to contribute to their sustainability. Moreover, the application of renewable energy technologies in the city no longer depends only on their technical evolution but on other factors and technologies that collaborate in improving their efficiency and performance. Some of these are summarised in table 13.1.

## 13.4 Final considerations

Therefore, when considering how to implement energy efficiency and RESs to contribute to more sustainable cities, it is clear to see that it is not possible to achieve this goal with only one technology; it requires the application of a mix of RESs. Moreover, not only are different technologies needed, but also different capacities and powers. This is the start of distributed generation.

The introduction of clean renewable energies will, in general, be conditioned by the level of technological development and economic efficiency, since it is in the electricity sector that the technological development of PV and wind generation equipment has been the most intense.

However, this energy transition will be very different between cities and between countries. In Europe, the European Community is embracing this transition with new policies and with economic support for the different initiatives for the decarbonisation of not only the energy sector, but also the building sector.

In a city, energy consumption is not only related to its dynamics or idiosyncrasies, but also to the progress of technology (i.e. development of light-emitting diodes in lighting) and to policies that embrace energy efficiency and its labelling.

The available interventions could be divided into a sequence in order of difficulty:

a. Evaluate renewable energy technologies appropriate to the size and geographic location of the city. Today, technologies are mature enough to be used, as access to them is possible. Roofs in cities give the possibility of implementing solar PV and solar thermal systems. Also, small-scale

**Table 13.1.** RES technologies for use in cities.

| Technology | Low power (1000 W—10 kW) urban/city | Medium power (10 kW—1000 kW) periurban | High power (>1000 kW) external |
|---|---|---|---|
| Solar photovoltaic (PV) | Household up to 4 kW Average 6.4 kW | Commercial roofs 50 kW–200 kW | Solar farms |
| Solar thermal | Household 2–4 $m^2$ 3 $kW_{th}$ | Hotels 20 $kW_{th}$ | Solar power plants (CSP) |
| Wind energy | Up to 4 kW vertical axis | Horizontal axis | Wind farms utility >5000 kW |
| Hydraulic | Use of slopes | Distributed tidal energy | Hydropower plant |
| Geothermal | Space heating | Space heating in commercial buildings space cooling | District power and water plants >30 MW |
| Biomass—thermo-conversion | Use of briquettes Pellets | Use of briquettes pellets | Urban waste incineration biorefineries |
| Biomass—bioconversion | Liquid biofuels Biogas | Biodigester for local use | Biorefineries |
| Flywheels | | Frequency-level distributed control | Grid-level frequency control |
| Pumping station | | Station for water control | Pumping station in mountains 300–1200 MW |
| Batteries | Energy storage | Energy storage frequency-level distributed control | Grid-level frequency control |
| Hydrogen | Household generation | Mobility fuel cell generation | Underground storage power plant |
| TES sensible | Household comfort | Large building acclimatisation (heat) | Seasonal energy storage |
| TES latent (PCM) | Household comfort | Large building acclimatisation (heat) | |
| TES thermochemical | Distributed heating-cooling (consumer) | Large building acclimatisation (heating/cooling) seasonal energy storage | |
| TES thermo-mechanical | | | Peak trimming seasonal energy storage |

aerothermal or geothermal energy should be considered. Both grid-connected systems and stand-alone ones are possible.

b. Stimulate local, regional, or national policies to facilitate RESs.
c. Stimulate investment by banks or energy service companies (ESCOs).

d. Stimulate local labour to increase RES deployment.
e. The other indirect beneficiaries of the application of these technologies are the same companies that provide electricity, which, by having small RES facilities, reduce the need to extend the grid or the damage caused by supply interruptions due to bad weather or poor-quality service techniques.
f. Distributed energy systems contribute to the resilience of the electricity grid.

According to the Global Sustainable Electricity Partnership, in terms of energy, cities are heading towards:

- The electrification of transport in cities, which will radically reduce local pollution and facilitate a digital revolution that will offer a wide range of new services to citizens. Clean and efficient electricity is the lifeblood of the modern city.
- New technologies that are much more efficient than the ones they replace: for example, electric heat pumps consume four times less energy than oil or gas boilers. In addition, they can be used for both space heating and air conditioning.
- Demand-side management and distributed renewable energy in buildings that can help to balance energy supply and demand and reduce costs for the system and for end-consumers.
- Encouraging the use of smart thermostats, electric boilers, and heat pumps that can also take advantage of the flexibility of customer demand instead of adding power generation and grid capacity.
- Digitisation that facilitates demand response to reduce load peaks and, by monitoring real-time performance of buildings, allows consumers and building managers to identify where and when maintenance is needed and where energy savings can be achieved.
- Residential energy storage systems that offer balancing services to the electricity grid through the use of PV systems.
- Thermal energy storage, which is a promising alternative in cities due to the possibility of using various alternative technologies at different power scales for the different spaces that cities and their surroundings may have.

In general, cities show remarkable potential within their immediate boundaries to implement one or more RESs. According to IRENA, regardless of whether they have set renewable energy targets, cities must assess these potentials both technically and economically, as the deployment of renewable energy brings more than just carriers of physical energy, but also many co-benefits.

> "*The market has given the signal with cost-competitive technologies. Policymakers must now establish the right frameworks to accelerate climate-proof investments.*
>
> *We must create a low-carbon energy system to keep the line in the face of rising global temperatures. It's possible*". — Francesco La Camera, Director-General International Renewable Energy Agency (IRENA 2019)

## Bibliography


Barragan-Escandon E, Zalamea E, Terrados J and Vanegas P 2019 Factores que influyen en la selección de energías renovables en la ciudad *Revista Latinoamericana de Estudios Urbano Regionales* **45** 259–77

Boshell F *et al* 2020 *Innovation Outlook: Thermal Energy Storage* (Abu Dhabi: International Renewable Energy Agency)

Bridgwater A V 1995 The technical and economic feasibility of biomass gasification for power generation *Fuel* **74** 631–53

Cabeza L F, de Gracia A, Zsembinszki G and Borri E 2021 Perspectives on thermal energy storage research *Energy* **231** 120943

ESA 2020 https://energystorage.org/why-energy-storage/technologies/

Gutiérrez G I Z, Carballo A E G and Richard N 2018 *Manual para la implementación de un Sistema de Gestión de la Energía en Gobiernos Municipales* (Bonn: Deutsche Gesellschaft für Internationale Zusammenarbeit (GIZ) GmbH)

Intedya 2017 La eficiencia energética de las ciudades inteligentes http://intedya.com/internacional/1296/noticia-la-eficiencia-energetica-de-las-ciudades-inteligentes.html

IRENA 2019 *Transforming the Energy System—And Holding the Line on the Rise of Global Temperatures* (Abu Dhabi: International Renewable Energy Agency)

Keirstead J and Shah N 2013 *Urban Energy Systems. An Integrated Approach* 1st edn (Routledge)

NETL 2020 National Energy Technology Laboratory 'Biomass' https://netl.doe.gov/research/coal/energy-systems/gasification/gasifipedia/biomass

Romero I, Navntoft C and Cristófalo M 2019 *Sistemas solares térmicos compactos para ACS. Dimensionamiento e Instalación* (Argentina: Secretaría de Gobierno de Energía) Ciudad Autónoma de Buenos Aires: Subsecretaría de Energías Renovables y Eficiencia Energética Edición

Salmeri M, Wedgwood P, Rugg R and Aiello R 2017 *Desarrollando soluciones de energías limpias en las principales ciudades de América Latina Introducción para los tomadores de decisión de política energética a nivel subnacional* (Carbon Trust and Banco Interamericano de Desarrollo (BID))

Sarbu I and Sebarchievici C 2018 A comprehensive review of thermal energy storage *Sustainability* **10** 191

Sharifia A and Yamagata Y 2015 A conceptual framework for assessment of urban energy resilience *Energy Procedia* **75** 2904–9

Shilston R and Danks R 2018 Simulation of urban solar reflections and their impact on building performance *CIBSE Technical Symp. (London, UK)*

Singh J 2016 Why Energy Efficiency Matters and How to Scale It Up *Live Wire; 2016/53* (Washington, DC: World Bank) License: CC BY 3.0 IGO https://openknowledge.worldbank.org/handle/10986/23636

Tarragona J, Pisello A L, Fernández C, de Gracia A and Cabeza L F 2021 Systematic review on model predictive control strategies applied to active thermal energy storage systems *Renew. Sustain. Energy Rev.* **149** 111385

Urge-Vorsatz D, Khosla R, Bernhardt R, Chan Y C, Vérez D, Hu S and Cabeza L F 2020 Advances toward a net-zero global building sector *Annu. Rev. Environ. Resour.* **45** 227–69

**IOP** Publishing

Sustainable Urban Development
Topics, trends and solutions
**Luís Bragança, Cristina Engel de Alvarez and Luisa F Cabeza**

# Chapter 14

# The role of materials in achieving sustainability in cities

**Luisa F Cabeza**

This chapter discusses the potential of materials to support the sustainability of cities. The increase in building floor areas worldwide means that a large quantity of materials is used. Several aspects are studied, such as the reduction of energy demand in buildings, the reduction of embodied energy and embodied carbon, the reduction of urban heat islands, and how to increase the contribution of materials in the circular economy. Finally, several recommendations are presented, such as the need to involve all stakeholders in the decisions, and the need for research.

## 14.1 Introduction

According to the International Energy Agency (www.iea.org), buildings and the construction sector together account for more than 30% of global final energy consumption and nearly 40% of total direct and indirect $CO_2$ emissions. One of the main drivers of buildings' energy demand is the growth in buildings' floor areas, that is, the increase of new buildings (Ürge-Vorsatz *et al* 2020). An increase in buildings' floor areas also means an increase in the quantity of materials used. Therefore, there is a need to decrease the materials used in the buildings in our cities and rural areas, but materials also impact other parts of the lifecycle of buildings, since the choice of materials impacts the embodied energy and embodied carbon in buildings, and also the operational stage of the lifecycle, since materials have a clear role in decreasing the energy demand of buildings.

Therefore, this chapter summarises the following aspects of the role of materials in the sustainability of cities: the reduction of energy demand in buildings, the reduction of embodied energy and embodied carbon, and the reduction of the urban heat island. Finally, some hints on how to increase the contribution of materials to the circular economy are given.

doi:10.1088/978-0-7503-3971-1ch14

## 14.2 Materials that can reduce energy demand in buildings

It is well known that it is necessary to insulate buildings to decrease the thermal demand and also to decrease the use of heating and cooling (Cabeza *et al* 2010). The first time a building was insulated was in the USA in 1880, when builders installed mineral wool in houses. From the 1970s, new and more effective insulation materials have been discovered. The long list of insulation materials includes the conventional ones (mineral wool, foam glass, perlite, vermiculite, cork, and cellulose), the more advanced ones such as aerogel and vacuum, and those known as the sustainable ones, which include bioinsulation materials (sheep's wool and agricultural waste) and recycled insulation materials. A summary of the properties of insulation materials is presented in table 14.1.

Another way to reduce the energy demand of buildings is to increase the thermal inertia using materials with a high thermal mass (Navarro *et al* 2016). High-thermal-mass materials are those that are able to store a large amount of heat. These properties provide thermal stability and smooth the thermal fluctuations between external and internal conditions. The high-thermal-inertia materials most used in buildings are stone, adobe, rammed earth, brick, water, and concrete.

Thermal inertia can also be added to the building by the incorporation of phase-change materials (PCM) in the building envelope (via latent thermal energy storage). During the solid–liquid phase change, the temperature remains constant, and the material can store large amounts of heat. PCM can be incorporated inside the building material itself, for example, in the cases of concrete or gypsum. Moreover, the PCM can be incorporated inside a construction element or macro-encapsulated

**Table 14.1.** Properties of insulation materials (adapted from Cabeza *et al* 2010).

| Type | Material | Density (kg $m^{-3}$) | Thermal conductivity (mW m·$K^{-1}$) | Specific heat capacity (J g·$K^{-1}$) | Vapor diffusion resistance | Sound absorption coefficient |
|---|---|---|---|---|---|---|
| Conventional | Rock wool | 40–200 | 33–40 | 0.8–1.0 | 1.0–1.3 | 0.29–0.9 |
| | Expanded polystyrene | 18–50 | 29–41 | 1.25 | 20–100 | 0.22–0.35 |
| | Polyurethane | 30–160 | 22–35 | 1.3–1.45 | 55–150 | — |
| | Perlite | 32–176 | 40–60 | 0.2 | 3.5 | 0.2–0.75 |
| | Vermiculite | 64–130 | 40–64 | 0.84–1.08 | 3–5 | 0.8 |
| | Cork | 100–120 | 37–43 | 1.5–1.7 | 5–30 | 0.39–0.85 |
| | Cellulose | 30–80 | 37–42 | 1.3–1.6 | 1.7–3.0 | 0.53–0.9 |
| Advanced | Aerogel | 70–150 | 13–21 | 1.0 | 2.0–5.5 | 0.54–0.78 |
| | Vacuum insulation panel | 160–230 | 3.5–8 | 0.8 | 340 000 | 0.1–0.3 |
| Sustainable | Bamboo fibres | 431–538 | 77–88 | 1.79–1.96 | 6.7–15.4 | 0.2–0.56 |
| | Rice husk | 130–170 | 48–80 | 1.2–2.7 | 2 | 0.15–0.66 |
| | Wood fibre | 50–270 | 38–50 | 1.9–2.1 | 1–5 | 0.1–0.32 |
| | Cotton fibres | 150–450 | 58–82 | 0.13 | — | 0.5–74 |

**Table 14.2.** Advantages, disadvantages, and properties of PCMs used to increase thermal inertia of buildings (adapted from Cabeza *et al* 2011).

| PCM type | Organic | Inorganic |
|---|---|---|
| Advantages | Non-corrosive | High phase-change enthalpy |
| | Low or no subcooling | Low cost |
| | Chemical and thermal stability | |
| Disadvantages | Low thermal conductivity | Corrosion |
| | Flammability | Phase segregation |
| Phase-change enthalpy | Around 120 J $g^{-1}$ | Up to 250 J $g^{-1}$ |
| Thermal conductivity | Up to 0.5 W $m^{-1}\cdot K^{-1}$ | Around 2 W $m^{-1}\cdot K^{-1}$ |
| Density | 500–1000 kg $m^{-3}$ | 900–1200 kg $m^{-3}$ |

and included as another layer. The efficiency of these elements depends on several factors, such as how the PCM is incorporated in the material, the orientation of the wall with the PCM, climatic conditions, direct solar gains, internal gains, ventilation rate, the PCM chosen and its phase-change temperature, the temperature range over which phase change occurs, and the latent heat capacity per unit area of the wall.

PCMs can be organic or inorganic (see the advantages and disadvantages in table 14.2), as hundreds of materials have been identified as suitable for use as PCMs (Cabeza *et al* 2011). The selection of the right material depends on its properties; its melting temperature is the one that ensures its suitability for a specific application. Some of the other very important properties are the phase-change enthalpy, lack of subcooling and lack of segregation, good compatibility with the encapsulation material, thermal conductivity, and cost. Typical values for PCM properties are summarized in table 14.2.

## 14.3 Materials to reduce embodied energy and embodied carbon in buildings

In 2018, 11% of $CO_2$ emissions were due to the manufacture of building materials and products, such as steel, cement, and glass (www.iea.org); this is known as embodied carbon in buildings. Due to the large amount of material used and the carbon-intense production processes, more than 50% of these emissions are attributed to steel and cement production. The other main materials used in buildings are wood (including bamboo), rammed earth (including adobe), and ceramics or masonry (including bricks) (Cabeza *et al* 2021). Other materials that contribute to embodied energy and embodied carbon in buildings are aluminium, glass, insulation, plastics, and copper.

It is recognized that the progress towards low-energy-demand buildings (such as net-zero energy buildings) increases the importance of embodied energy, due to the reduction in the operational energy demand (Pomponi and Moncaster 2016). Literature on the topic of embodied carbon is scarcer than for embodied energy, but both concepts have increased in interest recently.

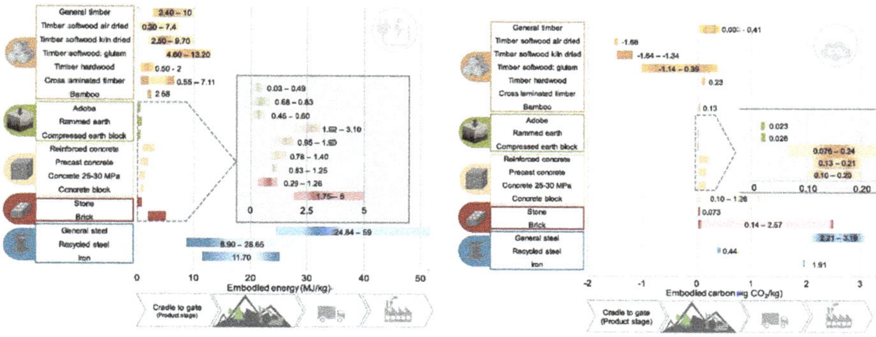


**Figure 14.1.** Cradle to grave embodied energy (left) and embodied carbon (right) of structural building materials (Cabeza *et al* 2021).

Cabeza *et al* (2021) carried out an homogenisation of both the embodied energy and the embodied carbon of structural building materials, which considered the product stage of the lifecycle of buildings (cradle to grave). The building structure construction materials identified were concrete, steel, masonry, rammed earth, and wood. The results are presented in figure 14.1.

The results show that the materials with the lowest embodied energy are earth materials (~0.60 MJ $kg^{-1}$), and that steel has a higher embodied energy (32–35 MJ $kg^{-1}$). Wood materials, including timber, have an impact of ~2.5–3.0 MJ $kg^{-1}$; for these materials, there is a large dispersion of values, depending on the type of wood (type of timber, glulam, laminated wood, etc). Similarly, earth materials are have a wide range of types (adobe, rammed earth, or compressed earth blocks), again giving variability to the embodied energy; but the addition of stabilizers, such as cement or natural fibres, also adds variability. The high embodied energy of concrete is attributed to the production of cement; the embodied energy for concrete is ~0.78–1 MJ $kg^{-1}$, and for reinforced concrete it is ~2–2.5 MJ $kg^{-1}$. Finally, the embodied energy for masonry is ~0.97 MJ $kg^{-1}$, but for brick, it is ~2.5–3 MJ $kg^{-1}$, due to the high temperatures required for its production.

The results for embodied carbon show large differences between the traditional materials (i.e. rammed earth, stone, timber), which have 0.2–0.4 kg $CO_2$ $kg^{-1}$, and steel (~2.5–2.7 kg $CO_2$ $kg^{-1}$). The embodied carbon of concrete is reported to be ~0.1–0.16 kg $CO_2$ $kg^{-1}$ and for reinforced concrete, it is ~0.19–0.24 kg $CO_2$ $kg^{-1}$, similar to that of bricks (~0.14–0.23 kg $CO_2$ $kg^{-1}$).

## 14.4 Materials for reducing the urban heat island

An urban heat island (UHI) is an urban or metropolitan area that is significantly warmer than the surrounding rural areas due to human activities. High population density, dense high-rise buildings, and impervious pavements aggravate UHI. One of the main consequences of UHI is an increase in the energy demand of buildings in cooling-dominant zones (about 10%–16%) (Santamouris 2014; Tian *et al* 2021).

The mitigation strategies for UHI are urban geometry, urban greening, water bodies, urban ventilation, and the use of cool materials. The use of light-coloured,

**Table 14.3.** Properties to be addressed for UHI mitigation and their improvement measures (adapted from Nwakaire *et al* 2020).

| Property | Description | Improvement measures |
|---|---|---|
| Albedo | – Proportion of incident solar radiation reflected by a surface<br>– Surfaces with lower albedo have higher temperatures<br>– Ranges between 0 and 1 | – Surfaces with light-coloured and smooth textured materials<br>– Surfaces with light-coloured paints<br>– Adding a reflective coating to surfaces |
| Emissivity | – The ratio of solar radiation emitted by the surface to the radiation emitted by a black body at the same temperature<br>– It is the most important factor affecting surface temperatures at night<br>– Ranges between 0 and 1 | |
| Absorptivity | – The ratio of the radiation absorbed to the radiation received by the surface<br>– Ideally, the albedo plus the absorptivity should be 1 | |
| Thermal conductivity | – A measure of the ability of a material to transfer heat<br>– Measured in W m·K$^{-1}$ | – Addition of conductive or insulating fillers<br>– Inclusion of graphite powder |

low-thermal conductivity materials has been developed for this purpose (Lai *et al* 2019; Santamouris and Yun 2020; Tian *et al* 2021). The mitigation strategies also include cool roofs, cool coatings for building surfaces, and cool pavements. The properties required to achieve UHI mitigation are summarized in table 14.3.

Cool roofs include the adoption of smooth and light-coloured roofing, painting existing roofs with white or light-coloured paints, or including greenery in the roof. Cool roofs increase reflectivity and rooftop albedo, improve storm water management, and save on energy and air conditioning costs (Nwakaire *et al* 2020).

The use of reflective surfaces substantially reduces the energy demand for cooling in buildings. The difference between a white coating (albedo of 0.72) and a black one (albedo of 0.08) can be up to 45 K on a clear warm day. While the cooling capacity of natural materials is significant, recent research has developed artificial materials that are able to achieve much lower surface temperatures and a considerably higher mitigation capacity (figure 14.2). The main characteristics of the material technologies developed are shown in table 14.4. Although coloured materials present a quite high reflectance and a relatively lower surface temperature than the dark coloured surfaces, advanced artificial white materials may reach surface temperatures that are as much as 15 °C lower than those of conventional materials. Photonic materials can

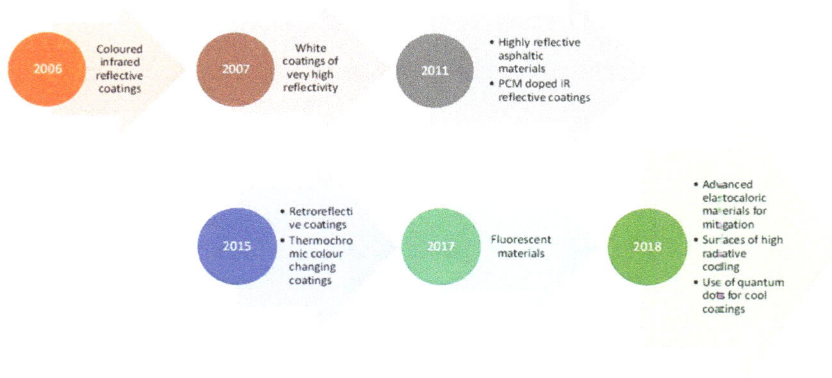


**Figure 14.2.** Development of new materials for UHI mitigation (adapted from Santamouris and Yun 2020).

**Table 14.4.** Optical properties of the materials technologies developed for UHI mitigation (adapted from Santamouris and Yun 2020).

| Type of material | High solar reflectance in the visible spectrum | High solar reflectance in the IR spectrum | High broadband emissivity | High emissivity in the atmospheric window | High fluorescent emission |
|---|---|---|---|---|---|
| Light-coloured reflective coatings | Yes | No | Yes | No | No |
| Coloured IR reflective materials | No | Yes | Yes | No | No |
| Reflective materials with nano-PCMs | Yes | Yes | Yes | No | No |
| Thermochromic materials | Yes | Yes | Yes | No | No |
| Fluorescent materials | Yes | Yes | Yes | No | Yes |
| Photonic materials and components for daytime radiative cooling | Yes | Yes | No | Yes | No |

achieve temperatures 5 °C–10 °C lower than cool white materials. Thermochromic materials and photonic materials can also substantially decrease the peak ambient and surface temperatures in buildings. However, these new materials have limitations and operational challenges, such as ageing and weathering problems. Therefore, further research to decrease these limitations has been carried out, such as adding phase-change materials (PCM) to polymeric cool roofs (Pisello *et al* 2016; Piselli *et al* 2019).

Pavements cover between 30% and 40% of urban areas and the temperature of pavements has been reported to be 20 °C higher than the temperature of the surrounding environment; therefore, strategies to reduce pavement temperatures are very interesting UHI mitigation strategies (Cheela *et al* 2021; Nwakaire *et al* 2020). Pavement properties such as albedo, thermal conductivity, and heat capacity are the ones that need to be modified, always considering that changing one of these influences the thermal equilibrium. Mitigation strategies include changing the pavement reflectivity.

## 14.5 Materials for increasing circularity in cities

Buildings are responsible for 40% of the materials consumed worldwide (Pomponi and Moncaster 2017); therefore, in the new context of the circular economy and better management of resources, buildings and the built environment need to include circularity in the use of materials. But buildings are complex entities, with long lifespans and changes of use during their service lives. Circularity in buildings can

**Table 14.5.** Examples of wastes from different origins used within construction materials (adapted from Smol *et al* 2015).

| Origin of wastes | Wastes | Construction materials |
|---|---|---|
| Agroindustry | Fly ash | Aggregate |
| | Bottom ash | Concrete |
| | Rice husk | Supplementary cementing materials |
| | Palm oil fuel ash | Blended cement |
| | Organic fibres | Bricks |
| | | Tiles |
| | | Blocks |
| | | Particleboard |
| | | Wall panels |
| | | Roof sheets |
| | | Reinforced polymer composites |
| Industry | Phosphogypsum | Fine and coarse aggregates |
| | Waste glass | Blended cement |
| | Granulated blast-furnace slag | Concrete |
| | Waste steel slag | Bricks |
| | Rubber tyres | Blocks |
| | | Tiles |
| | | Ceramic products |
| Mining | Quarry dust | Fine and coarse aggregates |
| | | Concrete |
| | | Bricks |
| | | Tiles |
| | | Blocks |
| | | Surface finishing materials |
| Construction sector | Construction and demolition debris (concrete rubble, tiles, waste bricks, etc.) | Fine and coarse aggregates |
| | | Concrete |
| | | Bricks |
| | | Blocks |
| | | Sub-base pavement materials |

only be achieved by including different aspects such as biomimetics in building design; resource efficiency and technological advancement; recycling, reuse, green supply chains, and waste reduction; and policy, people, and society.

Within this context, resource efficiency can only be achieved by using natural materials, and by recycling and reusing byproducts in buildings. Here, we include byproducts from the building sector itself, as well as from other sectors, such as agriculture. Examples of the use of agro-wastes in buildings are their use as insulating materials (see above), but also as composites within bricks, cementitious materials, or for the reinforcement of other materials (Madurwar *et al* 2013). Examples of wastes used in the production of construction materials are shown in table 14.5 (Smol *et al* 2015).

## 14.6 Conclusions and recommendations

Materials have a big role in increasing the sustainability of cities, which is due not only to the huge amount of materials used in cities, but also to their potential to increase/reduce the embodied energy and carbon, a very important aspect in today's challenge to mitigate climate change. Materials can reduce energy demand in buildings via insulation or by increasing thermal inertia. Materials can also reduce urban heat islands in urban areas. Finally, the circularity of cities can be increased by the use of natural materials and by recycling and reusing byproducts in buildings, pavements, and other urban areas.

But there is still a lot to do. Policymakers, designers, users, and all other stakeholders should always consider the potential of materials to increase sustainability in cities. There is a huge need to reduce the amount of materials used, but also to use recycled materials, and to ensure that materials can be reused/recycled after the lifespan of buildings, pavements, etc.

There is also a need for research to increase the materials reused and recyclability. Moreover, new materials, especially biobased materials, need to be developed with the appropriate properties (mechanical, physical, and chemical).

## Acknowledgements

This work was partially funded by the Ministerio de Ciencia, Innovación y Universidades de España (RTI2018–093849-B-C31—MCIU/AEI/FEDER, UE) and by the Ministerio de Ciencia, Innovación y Universidades—Agencia Estatal de Investigación (AEI) (RED2018–102431-T). The author would like to thank the Government of Catalonia for the quality accreditation given to her research group (2017 SGR 1537). GREiA has been certified as an agent TECNIO in the category of technology developers by the Government of Catalonia. This work is partially supported by ICREA under the ICREA Academia programme.

## Bibliography

Cabeza L F *et al* 2010 Experimental study on the performance of insulation materials in Mediterranean construction *Energy Build.* **42** 630–6

Cabeza L F *et al* 2011 Materials used as PCM in thermal energy storage in buildings: a review *Renew. Sustain. Energy Rev.* **15** 1675–95

Cabeza L F *et al* 2021 Embodied energy and embodied carbon of structural building materials: worldwide progress and barriers through literature map analysis *Energy Build.* **231** 110612

Cheela V R S *et al* 2021 Combating urban heat island effect—a review of reflective pavements and tree shading strategies *Buildings* **11** 93

Lai D *et al* 2019 A review of mitigating strategies to improve the thermal environment and thermal comfort in urban outdoor spaces *Sci. Total Environ.* **661** 337–53

Madurwar M V, Ralegaonkar R V and Mandavgane S A 2013 Application of agro-waste for sustainable construction materials: a review *Constr. Build. Mater.* **38** 872–8

Navarro L *et al* 2016 Thermal energy storage in building integrated thermal systems: a review. Part 2. Integration as passive system *Renew. Energy* **85** 1334–56

Nwakaire C M *et al* 2020 Urban heat island studies with emphasis on urban pavements: a review *Sustain. Cities Soc.* **63** 102476

Piselli C *et al* 2019 Cool roof impact on building energy need: the role of thermal insulation with varying climate conditions *Energies* **12** 3354

Pisello A L *et al* 2016 Innovative cool roofing membrane with integrated phase change materials: experimental characterization of morphological, thermal and optic-energy behavior *Energy Build.* **112** 40–8

Pomponi F and Moncaster A 2016 Embodied carbon mitigation and reduction in the built environment—What does the evidence say? *J. Environ. Manage.* **181** 687–700

Pomponi F and Moncaster A 2017 Circular economy for the built environment: a research framework *J. Clean. Prod.* **143** 710–18

Santamouris M 2014 On the energy impact of urban heat island and global warming on buildings *Energy Build.* **82** 100–13

Santamouris M and Yun G Y 2020 Recent development and research priorities on cool and super cool materials to mitigate urban heat island *Renew. Energy* **161** 792–807

Smol M *et al* 2015 The possible use of sewage sludge ash (SSA) in the construction industry as a way towards a circular economy *J. Clean. Prod.* **95** 45–54

Tian L *et al* 2021 Review on urban heat island in China: methods, its impact on buildings energy demand and mitigation strategies *Sustainability* **13** 1–31

Ürge-Vorsatz D *et al* 2020 Advances toward a net-zero global building sector *Annu. Rev. Environ. Resour.* **45** 227–69

# Chapter 15

# Practical strategies for embodying the circular economy concept in the construction sector

**Kamel Mohamed Rahla, Luís Bragança and Ricardo Mateus**

The construction industry has often been criticised for imposing heavy environmental damage on the planet by intensively consuming resources and generating waste. The concept of the circular economy (CE) has recently gained momentum among scholars; it stresses the need to transition towards healthier practices for the environment without compromising economic progress. This chapter will tackle a research gap that is often highlighted by academia and policymakers, which is the lack of practical strategies for implementing this concept in structures that are predominantly meant to be demolished when they reach the end of their lives. Globally, the paradigm aims to erase the concept of waste, relying on regenerative sources and keeping materials, components, and systems in use at their highest value for as long as possible. Even though the CE seems straightforward, translating this thinking to the building level might be difficult. This study sheds light on how to integrate the concept of the CE into buildings throughout their life stages, which, on a bigger scale, will ultimately lead to healthier, efficient, and more sustainable cities.

## 15.1 Introduction

There is no doubt that the building industry contributes significantly to economic prosperity and social welfare. Still, economic development has led to substantial environmental damage, accounting for 36% of greenhouse gas emissions and 40% of energy consumption (Hossain and Ng 2020). As a response, professionals, academia, and policymakers have started considering the concept of the CE as a suitable pathway for the building sector, especially given that the building sector has been listed as one of the five priority sectors in the newly released European CE action plan.

The CE is considered to be a more effective approach to attaining higher levels of sustainability and moving away from the basic linear economy, which consists of

doi:10.1088/978-0-7503-3971-1ch15

'take, make, use, dispose'. It encourages the reduction of raw material inputs and eliminates waste throughout the whole lifecycle to create a closed-loop system. Recent studies have focused on recycling construction and demolition waste (C&DW) with little consideration of the reuse of construction products themselves, which might lead to a reduction in the volume of reclaimed materials (Rahla *et al* 2021a). The avoidance of C&DW at the end of a building's life is enjoying increased adoption, but further studies must be conducted, particularly at the design stages, to ensure better end-of-life scenarios (Eberhardt *et al* 2020). While buildings are typically not designed for deconstruction, the concept of reusing their assets after the ends of their lives should be considered in order to achieve higher levels of performance with less input material and less waste generation. Other scholars have also stressed the significance of the building stock, which represents the most important part of the produced man-made wealth of industrialised societies and underlined the urgent need to decrease material use and increase the quality and durability of the stock as a resource.

In this context, this chapter will explore the background of the CE and the strategies and approaches that can be applied at the building level, from the design to the end-of-life stages, to provide clearer insights and shed light on current research gaps related to more efficient strategies for implementing the CE.

## 15.2 Theoretical background

The need for better management of natural resources that are becoming scarcer, along with dire environmental impacts, has urged academia and policymakers to think about a new pattern in order to move away from the current dominant economic model that has been adopted since the industrial revolution. The idea of the CE was refined throughout the last century and was finally widely acknowledged during the last decade. Nevertheless, the roots of the CE are still ambiguous and cannot precisely be pinned to a single study. One common agreement is that the CE is deeply rooted in numerous schools of thought (Rahla *et al* 2021b).

Together with Genevieve Reday, the industrial analyst Walter Stahel published a research report entitled 'The potential for substituting Manpower for Energy' which dealt with waste prevention through a new approach to production and processes for creating a closed-loop economy (Stahel and Reday-Mulvey 1981). The concept of the performance economy was further refined by Walter Stahel, who advocated the importance of product life extension and the importance of selling services instead of products, which is considered a pivotal feature of the CE.

The German chemist and visionary Michael Braungart and the American architect Bill McDonough developed the concept of 'cradle-to-cradle' two decades ago to describe a process wherein all the involved materials fall into two cycles: technical and biological. Their framework deals with design for effectiveness instead of design for efficiency to optimise the positive impacts of materials and components.

The concept of industrial ecology emerged as a potential response to resource scarcity and volatile material costs. This approach intends to create closed-loop strategies to use waste as an input to eliminate the notion of undesirable byproducts while promoting regenerative uses of resources. Consequently, wastes and resources stream in a circular way between the different ecosystem components, and renewable energy powers those cycles.

The American professor, John Tillman Lyle, proposed a challenge to his students back in the 1970s, which was to find a new approach to society that kept the same living standards while relying on renewable resources with no environmental damage. The idea of regenerative design focuses on developing buildings and cities in such a way that they regenerate ecosystems. This concept had been formulated earlier for agriculture, but professor Lyle expanded it to the entire social-ecological system to broaden its ability to enhance its potential.

The Chinese government started to show an increasing interest in the concept by implementing explicit policies back in 2002. Subsequently, research into the CE has increased rapidly, especially after the CE promotion law came into force in 2009 to stimulate cleaner production, develop eco-industrial parks, and cope with rapid urbanisation. In Europe, the creation of the Ellen MacArthur Foundation and the release of the action plan by the European Commission accelerated the transition towards the CE by promoting regenerative design, designing out waste and closed-loop systems to retain the value of materials circulating in the economy.

Initially, the CE-related Chinese literature outlined the concept of the three Rs: reduce, reuse, and recycle. The action 'reduce' refers to minimising inputs (raw materials) and outputs (waste). 'Reuse' is the operation of using a product again for the same purpose when it reaches the end of its life and 'recycle' is the process of recovering waste to manufacture a new product.

The three Rs principle was later extended to a nine Rs principle to encompass more actions. The first strategy stresses the need for wiser product manufacturing and includes three actions: refuse, rethink, and reduce. The second strategy encourages product lifetime extension and consists of: reuse, repair, refurbish, remanufacture, and repurpose. The last strategy comprises recycle and recover through incineration.

The Ellen MacArthur Foundation outlined three main principles for embracing the CE. First, to keep and upgrade natural resources by relying on renewable sources. Second, to keep materials and components at their highest value within the economy; and third, to adopt system thinking and design out waste (The Ellen MacArthur Foundation 2012).

Applying CE principles to buildings offers the promise of attenuating negative environmental impacts by keeping materials and components in use and guaranteeing their reuse when the building reaches the end of its life. Designing for a circular economy implies designing for adaptability, flexibility, and disassembly to enable reversibility and salvage the value of a building's products. This decreases the use of raw materials and reduces waste generation from the design stage onward.

Nevertheless, merging CE principles into the built environment is currently hampered by the inherent complexity of the concept and the lack of general consensus surrounding it. Additionally, buildings are still predominantly constructed to be demolished at the end of their lives, instead of deconstructing them or adapting them to users' needs.

## 15.3 Circular economy strategies for resources

Buildings are well-established as major consumers of raw materials, energy, and water. By adopting particular CE strategies, savings may occur throughout the whole of a building's lifecycle. A circular approach to buildings is to close material, component, energy, and water loops and minimise the associated environmental impacts. Hence, the lifecycle of circular buildings should be a closed-loop system wherein components and materials are optimally used and retained at their highest value. The technical cycle consists of selecting materials and components that can be maintained to extend their service lives by reuse, refurbishment, repair, or remanufacture, whereas recycling and incineration are considered to be the ultimate strategies. Alternatively, the biological cycle involves natural materials that can be biodegraded or composted at the end of their lives.

As highlighted by Pomponi and Moncaster (2016), circular buildings encompass green and sustainability strategies. The environmental footprints of materials and components should be as minimal as possible. It is preferable to select materials that are sourced locally to reduce the emissions due to transportation and stimulate the local economy. Low embodied energy and carbon materials that are abundant, renewable, and pure are fundamental strategies for achieving eco-friendly structures with a minimal input of raw materials within a circular flow system. Since circular building design entails adaptability, flexibility, and deconstruction at its core, materials and components need to endure numerous life cycles, depending on their use purpose. Therefore, durability, resilience, and high quality are features that need to be taken into account when selecting materials and components to prevent the loss of quality or value over time and use cycles.

Energy is becoming a common theme among policymakers, given the potential positive impacts of relying on renewable sources. Recent policies for nearly zero-energy buildings are challenging practitioners and professionals to produce buildings with high energy efficiency. Circular buildings can achieve energy neutrality and even produce an excess of energy through design and local and renewable energy using smart grids. The overall energy consumption of buildings is mainly allocated to building operations, followed by embodied energy, which suggests that the use of materials and components that are energy-intensive increases the building's embodied energy. In Europe, the energy consumption of the building products industry ranges from 5% to 10% (European Commission 2014). Other key strategies to reduce energy consumption during the use phase are to consider the local climate for thermal comfort (e.g. enhance thermal insulation in cold areas or increase ventilation in hot ones), to implement architectural features to minimise artificial lighting and maximise the passive performance of the building, and to rely on on-site renewable energy to decrease the consumption of non-renewable energy. Moreover, high-quality materials are needed to provide the necessary services and thermal comfort for the users. Therefore, the selection of materials, contextual building design, and layout should be considered to reduce energy consumption throughout the building's lifecycle.

As a result of global urbanisation, water is getting scarcer, as this resource is encountering increasing demand many have predicted a fresh water crisis by the end of this century. In a building's lifecycle, water is extensively used and consumed, from material extraction to building demolition. Like energy consumption, water is most consumed at the operational stage. Water has numerous uses in the manufacture of building materials, such as: cement-based materials, ceramics, and steel. Selecting low-embodied-water materials will eventually lower the building's water footprint. A strategic approach to reducing operational water consumption is to adapt buildings to the context of their location to embrace more water efficiency. Water efficiency can be achieved during the use phase of circular buildings by harvesting rainwater and reusing grey water.

## 15.4 Design for the CE

One important aspect of CE thinking is to provide building materials, components, and systems a second life when the building itself reaches the end of its life. To enable reversibility, adaptability, and flexibility, a specific set of design strategies should be followed right from the early design phases, using a holistic approach. This section will cover different design strategies that could be put under the umbrella of designing for the CE (DfCE).

In recent decades, several studies have highlighted the multiple benefits of building disassembly (Crowther 2005; Durmisevic and Brouwer 2006). These studies were driven by the environmental and economic benefits resulting from designing for further reuse and recycling to cut down demolition costs and slow down the pressure on resource consumption. However, with the emergence of the CE, these practices gained more attention, as they matched the circular thinking to create a closed-loop resource flow.

Crowther (2005) outlined a total of 27 principles for Design for Assembly (DfD) that can be categorised as follows: reduce the amount of input and rely on safe and healthy secondary materials, standardise the connections between materials and components while considering disassembly at the end of life, and retain information regarding all materials and components involved in a building (Crowther 2005). Similarly, Ciarimboli and Guy (2005) framed DfD principles around the ease of deconstruction of building elements, the eradication of chemical links, reliance on mechanical connections that are accessible, dry construction, the provision of guidelines to support safe deconstruction, and the selection of suitable materials for the process.

One of the most crucial parameters in DfD is the choice of materials and components and how they will be assembled so that they can eventually be disassembled. Materials should embrace CE principles such as: reuse, refurbish, repair with higher purity to limit quality loss during the processes of assembly/disassembly. Unlike steel and timber, concrete has been overlooked when it comes to DfD, given the numerous challenges of disassembling concrete elements. Off-site construction and modularity are important features to consider when assembling components. The work of Brandt in his book 'Building in layers' has been fundamental to current strategies for DfCE (Brand 1995). In this book, the author described buildings as separate but somewhat interlinked layers, with their own technical and functional lifespans. Brand's widely-known model includes six layers with different lifespans (Brand 1995):

1. Site: the fixed location of the building (eternal);
2. Structure: the skeleton of the building, including the foundation and the load-bearing elements (30–300 years);
3. Skin: the façade, roof, and other building elements in contact with the external environment (20 years);
4. Services: the pipes, cabling, energy, and heating systems (7–15 years);
5. Space: the solid internal fit-out, including walls and floors (3–30 years);
6. Stuff: the rest of the internal fit-out, including the furniture and lightning (1 day–1 month).

Adaptability is another characteristic that that has been linked to DfCE. Addis and Schouten (2004) stressed the difference between adaptability and flexibility and defined a flexible building as '*a building that has been designed to allow easy rearrangement of its internal fit-out and arrangement to suit the changing needs of occupants*', whereas an adaptable building is '*a building that has been designed with thought of how it might be easily altered to prolong its life*'. Adaptable buildings should be maintainable, versatile, simple in design, and upgradeable to accommodate the users' desires with minimum quality loss and environmental impacts.

Similarly, disassembling sections and parts from buildings into components and allowing their reassembly in a new combination is a better alternative than destroying buildings and systems. Three dimensions of transformation were observed by Durmisevic and Brouwer (2006): spatial transformation, structural transformation, and element and material transformation.

**Table 15.1.** Strategies for embedding DfCE.

| Design approach | CE strategies |
|---|---|
| Design out waste | – Planned material reuse and recovery<br>– Off-site construction<br>– Optimised material use<br>– Lean design |
| Design for adaptability | – High interdependency between systems<br>– Components and systems with compatible service lives<br>– Upgradeable components and systems<br>– Modularity of building elements<br>– Building elements that can support multiple uses<br>– Transparency |
| Material selection | – Low embodied energy and carbon<br>– Reusable and recyclable materials<br>– Materials that are easy to disassemble<br>– Materials that can be put back into the technical or biological cycle<br>– Non-toxic and non-hazardous materials |
| Design for disassembly | – Safe and accessible disassembly<br>– Guidelines for deconstruction<br>– Standardised disassembly using simple tools<br>– Minimal chemical connections |

To adopt a DfCE strategy, practitioners and academia need to learn from the past to design for the future, as the CE is deeply rooted in several schools of thought and consequently accommodates numerous design strategies (table 15.1). It is necessary for designers to consider resource flows in building projects. As discussed in the previous section, an orderly selection of materials with low embodied energy and carbon that require minimal water consumption throughout their lifecycle is crucial to achieve positive environmental impacts. Water and energy should be considered during the design phase to produce buildings that optimise resource flows without compromising the overall quality and the users' comfort.

## 15.5 Digital practices in circular buildings

Information management is becoming an essential approach for the construction industry in order to manage the performances of buildings throughout their entire lifecycle, from the design stages until the end of life. Specific data are required to support the maintenance of a building and guarantee the technical life of all in-use materials. The tracking of resource flows allows building stakeholders to get deeper insights into the inputs and outputs of resources. Comprehensive and accurate data can enhance decision-making and manage financial risk. Digitising the built environment is regarded as a fast-track approach to implementing the CE in the building sector. Keeping track of materials through specific data sets supports materials' reuse once they reach the ends of their lives, which helps value retention. The information related to materials and components can be stored in a 'material passport' (MP) that can be shared with all the building's stakeholders.

The EU-funded project 'Buildings As Materials Banks' (wherein the authors participated, along with other prominent European institutions) was the starting point for the development of MPs. The objective of this project was to introduce the CE into buildings and develop digital tools to implement the concept for these assets while increasing the value of construction materials and components. The MP framework intends to support circular building design and information storage and sharing between all stakeholders. Comprehensive and accurate data can enhance decision-making, manage financial risk, and monitor environmental impacts. The MP provides both qualitative and quantitative information regarding the physical, chemical, and biological properties of a material's composition, and therefore allows practitioners to select building products that are safe for humans and ecosystems whilst taking into account CE principles.

'Building information modelling' (BIM) is a computational methodology that digitally models buildings to provide the construction stakeholders with valuable information that aids decision-making throughout the building's lifecycle. In the context of the CE, BIM can be considered as a way of increasing collaboration and decreasing waste generation. However, BIM is seldomly used for managing buildings when they reach the ends of their lives. Recent studies have attempted to link MP to BIM to compile data on building materials' characteristics and assess their performances. However, research supporting the integration of MPs into BIM is still in its early stages and requires deeper commitment from the construction industry. Digitising the building industry will still face several obstacles, such as the quality and the reliability of the used data and will require standardisation and transparency from the building stakeholders along the value chain.

## 15.6 Examples of CE application in construction projects

One of the first pilot projects to design, construct, and plan the deconstruction of a building was the project ‘Circular Building’ in London, developed by Arup during the London Design Festival in 2016. The designers of the project relied on the Brand 6S model to create a modular building with independent systems and mechanical connections. All the building products were obtained from sustainable sources and demonstrated flexibility and adaptability, while being able to be reused or recycled at their end of service lives. These materials are healthy both to humans and the environment and have low embodied carbon and energy. The innovative aspect of this project is the fact that the building is equipped with sensors to generate valuable data to control and adjust the quality of the indoor environment, from lighting and temperature to air quality.

Another interesting case study is the renovation of the headquarters of the energy grid company Alliander, located in Duiven, The Netherlands, which took place in 2015. The designers relied on MPs to store valuable information about the origins of the construction materials and products and their future reuse. Additionally, the building is energy positive, which means that it generates more energy than it requires through solar panels and underground water for thermal storage. The project used a minimum of raw materials, upgraded materials already in use, and recycled wood and steel.

In 2013, the first building project to use an MP was the Brummen Town Hall, which received, in the same year, the Dutch Award for Sustainable Architecture. The building, which was designed for future disassembly, extended a historic structure from the 19th century that was restored beforehand. The materials used were in line with CE principles and can be further dismantled to be relocated and used in a new structure. As the building was designed for a service life of 20 years, the materials passport plays a key role in this project, keeping track of the materials and components used in order to facilitate their reuse.

On a larger scale, the Circle House Project is considered to be the first social housing project to apply CE principles to keep materials at a higher value. The Danish project started in 2019, and the aim was to construct 60 circular social houses with reusable and recyclable construction materials. The participation of 30 companies in the entire construction process has increased collaboration among all stakeholders across the value chain. The Circle House Project will allow the construction industry to learn from this experience while developing adequate business models.

## 15.7 Concluding remarks

With the emergence of the CE, construction stakeholders are considering the adoption of the paradigm and the implementation of its principles in their building practices. This will require a paradigm shift towards sustainable resource

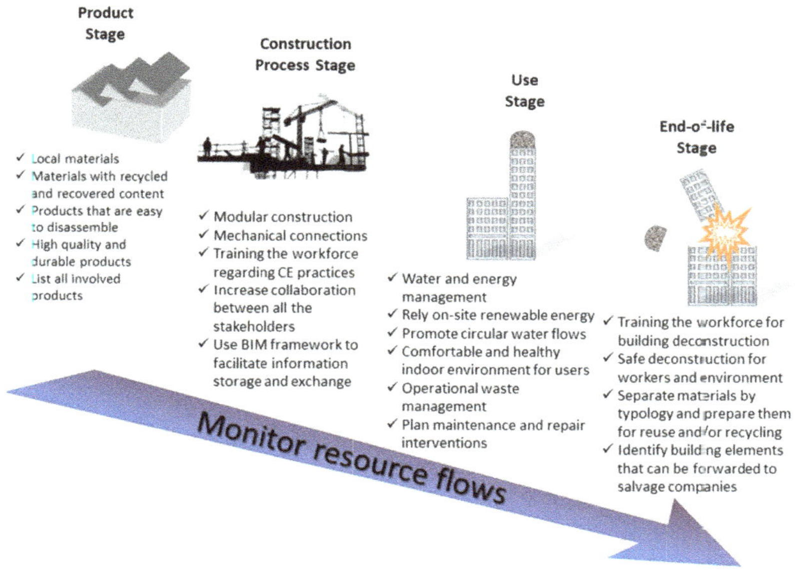


**Figure 15.1.** CE implementation strategies throughout the stages of a building's life.

management and a predictable end-of-life scenario for buildings, their materials, and their components (figure 15.1).

Adopting CE principles in the built environment will allow a reduction of the consumption of resources and the production of waste, retain the value of resources as long as possible within the system, and reintroduce resources into the use phase through particular strategies. One important approach to embedding circularity is to select suitable construction materials with low embodied energy and low embodied carbon that can enable deconstruction and other design strategies to be used, in line with the CE principles. Circular building design should encompass low energy and water consumption throughout its lifecycle. Energy efficiency practices and local and renewable energy sources are valuable approaches to consider, along with CE principles, to attain energy neutrality. Similarly, water consumption can be minimised through an adequate building design and reuse and harvesting systems for rainwater and wastewater. To adjust resource flows and reach higher levels of resource efficiency, a valuable approach would be to monitor resource consumption in the operational phase, which can provide relevant insights to buildings actors.

In the past two decades, several studies have highlighted the need for design strategies that can enable material reusability and recyclability. With the increasing interest that the CE is gaining, these particular design strategies have been encapsulated under the paradigm's principles, as they matched the same objectives of value retention throughout several uses. To ensure the quality of materials and components and describe their specific characteristics, several studies have proposed the integration of MPs with building information management, which will allow building stakeholders to track materials and assess their quality.

Numerous case studies across the globe have attempted to concretise CE principles and have provided real-life examples of how the construction industry can apply this paradigm shift. However, along the way, several challenges were noticed. For instance, selecting the right building materials and components proved to be crucial for designers, in order to embed circularity along the value chain. Nonetheless, building professionals encountered a lack of collaboration and obstacles to the collection of relevant data from material suppliers. Furthermore, this type of innovative project requires close management to keep the expected objectives aligned. Workers, clients, team members and other building stakeholders will need to have a different mindset to embed circularity throughout the project lifecycle.

Another critical obstacle to the CE uptake is the lack of governmental support through legal frameworks to compel the industry to rely more on secondary raw materials and to take accountability for the end-of-life scenarios of construction products. It is still not quite clear whether adopting CE strategies could bring about positive impacts at larger scales of the built environment. Despite the increasing interest that the concept is gaining, further accurate and contextual research is needed to objectively quantify the environmental, economic, and social benefits of the CE.

Finally, the CE aims to redefine the construction industry and challenges designers to reconsider the way they design buildings. Although various pilot projects have gone beyond recycling to apply CE core principles, several obstacles still remain along the way, which will eventually delay the transition towards a circular built environment.

## Bibliography

Addis W and Schouten J 2004 *Principles of Design for Deconstruction to Facilitate Reuse and Recycling* C607 (London, UK: Construction Industry Research & Information Association (CIRIA))

Brand S 1995 *How Buildings Learn: What Happens After they're Built* (London, UK: Penguin)

Ciarimboli N and Guy B 2005 Design for Disassembly in the built environment: a guide to cloosed-loop design and building (Pennsylvania State Univ. State Coll. PA)

Crowther P 2005 *Design for disassembly–themes and principles* Royal Australian Institute of Architects 1–7 Environment Design Guide

Durmisevic E and Brouwer P J 2006 *Design aspects of decomposable building* International Council for Research and Innovation in Building Construction (CIB) 1–23 *Proc. CIB Task Gr. 39-Deconstruction Meeting.* F Schultmann, AR Chini, eds.

Eberhardt L C M, Birkved M and Birgisdottir H 2020 Building design and construction strategies for a circular economy *Archit. Eng. Des. Manag.* **0** 1–21

European Commission 2014 Resource Efficiency Opportunities in the Building Sector 445 Final https://ec.europa.eu/environment/eussd/pdf/SustainableBuildingsCommunication.pdf

Hossain M U and Ng S T 2020 Strategies for enhancing the accuracy of evaluation and sustainability performance of building *J. Environ. Manage.* **261** 110230

Pomponi F and Moncaster A 2016 Circular economy for the built environment: a research framework *J. Clean. Prod.* **143** 710–18

Rahla K M, Mateus R and Bragança L 2021a Selection criteria for building materials and components in line with the circular economy principles in the built environment—a review of current trends *Infrastructures* **6** 49

Rahla K M, Mateus R and Bragança L 2021b Implementing circular economy strategies in buildings—from theory to practice *Appl. Syst. Innov.* **4** 26

Stahel W R and Reday-Mulvey G 1981 *Jobs for Tomorrow: The Potential for Substituting Manpower for Energy* (New York: Vantage Press)

The Ellen MacArthur Foundation 2012 Towards a circular economy—economic and business rationale for an accelerated transition *Greener Manag. Int.* 97 https://emf.thirdlight.com/link/x8ay372a3r11-k6775n/@/preview/1?o

**IOP** Publishing

Sustainable Urban Development

Topics, trends and solutions

**Luís Bragança, Cristina Engel de Alvarez and Luisa F Cabeza**

# Chapter 16

# Integrated actions to address the COVID-19 pandemic and climate change impacts

**Nils Larsson**

## 16.1 Introduction

All countries around the world are now facing two major crises: the impacts of the COVID-19 pandemic and the effects of climate change.

Global population health is the most direct victim of COVID-19, but the pandemic has also had major negative impacts on global and national trade, economies, and travel.

Climate change is now recognized by most governments as a phenomenon that is creating disaster on many fronts, although impacts vary in degree and pace in different regions. The impacts and massive costs of climate change adaptation and mitigation are becoming apparent and will rapidly increase. Table 16.1 provides some broad points of comparison.

Human civilization truly finds itself between a rock and a hard place. However, the urgent need to address both these crises offers great opportunities for synergies that can be achieved by integrating post-pandemic goals with those of climate action.

## 16.2 The COVID-19 pandemic

Almost all public health authorities around the world are currently focussed on COVID-19, which is a 'novel' respiratory coronavirus that emerged in 2019. However, there have been many epidemics and pandemics in the past, several of which have had profound impacts on human civilization and events, and there is reason to think that this pattern will continue.

Figure 16.1 below shows recent updates on the situation and are based on the World Health Organization Coronavirus (COVID-19) Dashboard which provides updated information on a daily basis.

doi:10.1088/978-0-7503-3971-1ch16  

**Table 16.1.** An overview and comparison of the two crises.

| Factors | COVID-19 pandemic | Climate change (see section 16.3 for details) |
|---|---|---|
| Causes | COVID-19 is the human disease caused by the SARS-CoV-2 zoonotic virus, which probably originated in bats. COVID-19 is now considered to be primarily airborne in spread. | Main drivers of anthropogenic climate change are greenhouse gases produced by use of fossil fuels in industrial processes, electrical generation, transportation, buildings, agricultural processes, or land-use change; all driven by population size, industrialization, household consumption, and waste levels. |
| Main features | Primarily airborne virus with person-to-person spread, resulting in illness, mortality, and long-term secondary effects. Several variants with higher rates of transmission have emerged during 2020/21. | Primary effects of high levels of atmospheric greenhouses gas (GHG) include increased ambient temperatures, which is a main driver of many secondary and tertiary effects. |
| Extent | The spread of COVID-19 depends on national measures, such as travel restrictions and/or cultural patterns and individual behaviour. | All regions are being affected, but regional impacts are uneven since they are moderated and modified by thermal storage in oceans and major weather patterns in different latitudes. |
| Intensity of impacts | Human health impacts ranging from mild illness to death, also serious economic and psychological impacts. | |
| Speed | Rapid initial spread with global spread over a 12-month period. | Climate change began slowly with the Industrial Revolution in the late 19th century but is currently rapidly accelerating. |
| Duration | Most pandemics have cyclical outbreaks with a total duration of less than 3 years. | Anthropogenic climate change is a once-in-an epoch event that may end with the collapse of industrial civilization. |
| Counter-measures | Personal separation, quarantine, or isolation of households, vaccination programs. | Reduce energy consumption, eliminate use of fossil fuels, shift to renewable energy, and reduce consumption and waste at all levels. |

The COVID-19 pandemic has been and remains a global health, social, and economic disaster. A vaccination program is not likely to be globally implemented until 2022, and other novel viruses are also likely to emerge in the near future.

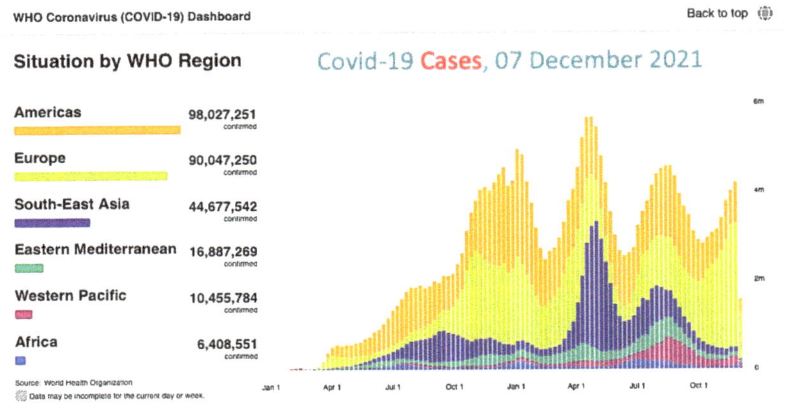


(a)

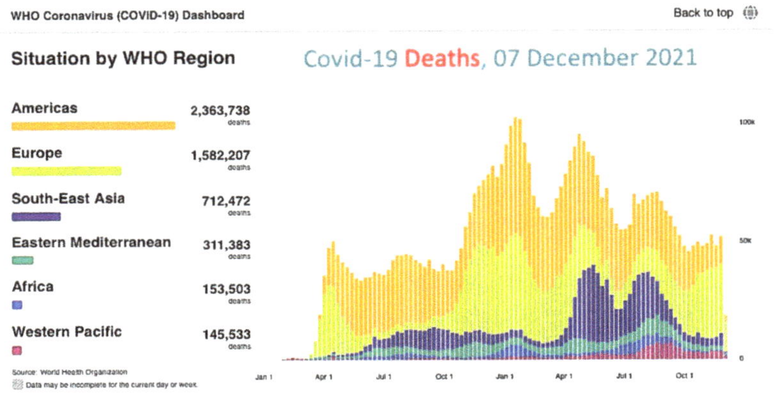


(b)

**Figure 16.1.** COVID-19 cases (a) and deaths (b) as of 30 April 2021, adapted from the World Health Organization COVID-19 Dashboard.

### 16.2.1 Highlights of COVID-19 impacts

Some highlights of the COVID-19 impacts are:

- Major negative effects on population health
- Overwhelming and uneven demand for health services
- Declining demand for urban office space and small retail space
- Major negative impacts on the 'real' economy, as evidenced by many consumer retail shops and services being forced to close down during cyclical COVID-19 outbreaks
- Increased teleworking and schooling at home
- Need to renovate public and educational buildings for COVID-19 needs
- Drastic reduction in demand for public transit
- Population movement towards suburbs and fringe communities in developed countries

- Increasing demand and prices for single-family houses in developed countries
- Loss of women's work opportunities gained over the last decades
- Increasing socio-economic inequalities

### 16.2.2 COVID-19 impacts on living and working

The need for spatial distancing and the limits applied to maximum group sizes in educational, work, or social settings have had multiple secondary effects.

The pandemic caused major changes, triggered by the almost-immediate and universal adoption of a 2 m distance between individuals to minimize the possibility of viral transmission. This criterion has led to a drastic reduction in maximum occupant densities in all occupied buildings and the closure of some types, such as bars and restaurants, depending on the stage of the pandemic cycle. Very recent statements by CDC now indicate a greater recognition of the role of airborne infection (www.cdc.gov/coronavirus).

Office work now involves greatly increased telework at home, with no, or very few, on-site meetings, while hourly-paid staff in hospitality facilities and small retail shops have lost jobs on a massive scale.

The degree of compatibility of some occupations with telework can be seen in figure 16.2. These factors have exacerbated socio-economic inequities between salaried office workers and hourly wage earners.

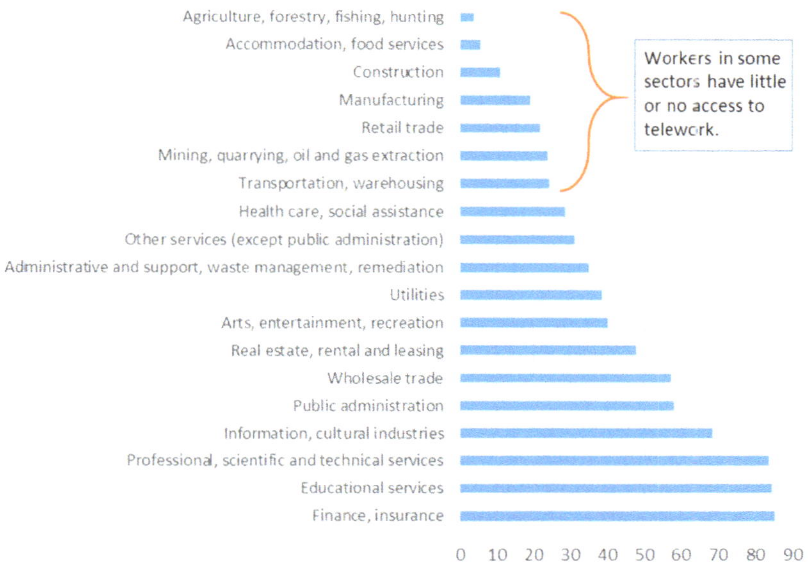


**Figure 16.2.** Telework capacity by industrial sector, Canada 2019.

In Western countries, schools have been closed for long periods during peak pandemic cycles, and many students have been asked to learn from home via computers and the internet[1].

### 16.2.3 COVID-19 and buildings

*a. Spatial distancing*

The need for spatial distancing has impacts on several common building elements:

- The most obvious impact on buildings is the reduced number of occupants in functional areas of buildings, based on density and distancing needs. In the case of office space, declining demand caused by reduced occupancy densities has been reinforced by the trend towards home teleworking (see (2) above).
- Schools and universities have also been affected by distancing requirements; remote learning is a solution, although the quality of education may suffer.
- Wide corridors are required to allow people to pass each other at a distance. This has affected supermarkets and large retail shops.
- Larger washrooms or, where enlargement is not feasible, reduced occupancy.
- Elevators must operate at reduced occupancy, which affects service capacity.

*b. Ventilation*

Other building-related impacts relate to the need for uncontaminated ventilation air, which is important because the main vector of spread for the COV-2 virus appears to be airborne. This is also related to physical distancing and face-mask requirements, which are based on preventing the transmission of airborne virus.

In the *ASHRAE Position Document on Infectious Aerosols* (ASHRAE 2020), the following statement is relevant: *Transmission of SARS-CoV-2 through the air is sufficiently likely that airborne exposure to the virus should be controlled.... Ventilation and filtration provided by heating, ventilating, and air-conditioning systems can reduce the airborne concentration of SARS-CoV-2 and thus the risk of transmission through the air.* Therefore, the most direct and effective way of ensuring ventilation air free of contaminants is to have operable windows to introduce outdoor air, if the outdoor pollution is minimal, or to ensure a high percentage of outdoor air in mechanical ventilation systems.

*c. Building types*

The COVID-19 pandemic has revealed important conflicts between pandemic health requirements and the ability for these to be satisfied in buildings of various occupancy types. Some requirements can be met using changes to building management or with minor renovations, but others demand more substantial changes that may not be easy to implement in the short term:

- Some building and occupancy types, such as long-term care facilities, have proven to be very unhealthy environments for the elderly occupants and their careers,

[1] Leaving aside issues of the quality of learning in such a situation, this causes major problems when home conditions are cramped and internet connections are poor or expensive. Home schooling requires supervision by an adult and, in almost all cases, it is the woman who is expected to take on this role, which may require abandoning paid employment.

mainly because of overcrowded rooms and poor ventilation. These factors have resulted in high mortality rates of both staff and residents. In Canada, 82% of all COVID-19 deaths during the March–April 2020 period were at long-term care facilities (data from National Institute on Aging, www.nia-ryerson.ca).
- Food processing plants have also proven to be unsuitable environments because of high occupancy densities and low interior temperatures that favor the survival of the COV-2 virus (Mohan *et al* 2021).

*d. Implication for building modifications or renovations*

Many other buildings will need modifications or renovations to make them pandemic-resistant. Issues include:
- Lifts in existing commercial or public office buildings have peak usage times that will require scheduling or an acceptance of reduced service. However, since many staff will be working at home, the problem is not likely to be a major one.
- Lifts in existing high-rise multi-family residential buildings pose a special problem if infection by the airborne COV-2 virus is to be prevented, because of the normally limited passenger capacity in such buildings. It is not possible to impose scheduled operations in such buildings, nor is physical enlargement usually possible.
- For all lifts, ventilation systems are likely to need replacement or modification to increase air change rates in such confined spaces.
- Circulation spaces in public buildings (including lobbies) do not appear to present special problems.
- Public washrooms are likely to need changes to limit occupancy to safe levels.

### 16.2.4 COVID-19 impacts on urban transport and mobility

Public transit systems are a major element in controlling emissions from urban transport, but the COVID-19 pandemic has had a drastic effect on ridership in public transit systems.

In three major international cities, the use of urban public transport systems decreased by 70% to 80% during the mid-February to early April 2020 period and has only partly climbed back. Data from the *American Public Transit Association* (https://transitapp.com/coronavirus) for March 30, 2021 from Paris, New York, and Montreal indicates that ridership on public transit systems fell by 10%, 53%, and 60%, respectively, compared to pre-pandemic levels during the last year. These drastic declines will cause the financial viability of such systems to be precarious unless measures are rapidly taken to rebuild confidence in these systems.

The shift away from public transport has resulted in an increased use of private vehicles, which has ominous implications for transport-related emissions in urban areas.

### 16.2.5 COVID-19 impacts and urban form

*a. Move away from central areas*

Massive changes in work and formal education patterns are causing an outflow of population to suburban and exurban areas of developed countries. This is having the unfortunate effect of reinforcing the role of cars as the main mode of transport, with negative effects on air quality and GHG emissions.

*b. Pandemic redundancies*

Parts of urban areas and buildings have been made redundant by the pandemic:

- Unused open parking spaces.
- Commercial retail
- Surplus office building space
- Surplus space in other building types, e.g., conference centres, gymnasia, shopping centres, parking garages, or parts of airports.

The pandemic has caused these facilities to fully or partially lose their original functions. The best use should be made of these opportunities to create new green spaces, playgrounds, urban agriculture, low-income housing, public transport, or other facility types that are suited to post-pandemic uses.

### 16.2.6 Post-pandemic recovery

*a. A 2020 survey of central bank and ministry officials and other financial experts*

A paper summarizing a survey of 231 central bank officials, finance ministry officials, and other experts from G20 countries (Hepburn *et al* 2020), suggests that the pandemic has … *precipitated a major increase in the role of the state* … and that … *the crisis has also demonstrated that governments can intervene decisively once the scale of an emergency is clear and public support is present* …

*b. IEA and IMF World Energy Outlook Special report*

The *World Energy Outlook Special report*, published in June 2020 by the International Energy Agency (IEA) and the International Monetary Fund (IMF) outlines their views on the prospects for an economic recovery (www.iea.org/reports/world-energy-outlook-2020):

- Regarding buildings…*More than 25 million jobs across the sector have been lost or are at risk in 2020. However, measures to improve the efficiency of buildings and appliances could be implemented quickly, in some cases with very short payback periods, creating 10–15 jobs per million dollars invested.*
- *Average annual energy retrofit rates in buildings are currently less than 1% in most major markets, which is well below the level required to achieve sustainability objectives.*

Policy approaches for addressing the current COVID-19 crisis suggested in the *World Energy Outlook special report* include:

- *Target efficiency improvement measures on* … *low-income households, small businesses and hotels.*
- *Use public procurement to catalyze activity, for example by commissioning efficiency retrofits of public assets such as social housing, schools, offices and healthcare facilities.*
- *Provide guarantees to encourage energy service companies to invest in retrofits.*
- *Accelerate or expand existing and planned efficiency programmes.*

*c. COVID-19 recovery plan by BPIE*

The Buildings Performance Institute Europe (www.bpie.eu) has carried out an analysis of the investment opportunity for Europe's building sector, which could help to mitigate the damage caused by the COVID-19 pandemic. According to BPIE…*a significant push for serial renovation is necessary to increase speed and depth of renovation while keeping the required work force at a realistic level. Industrial/serial renovation delivers cost-effective net-zero renovation at high quality by making use of prefabricated building modules, innovative financing and business models, and digitalization across the value chain. The paper presents an estimate to contribute to the discussion and to support decision-making.*

*All figures presented in the paper are related to either medium or deep renovation so that renovation efforts are aligned to the requirements of the Paris Agreement. The total amount to trigger a significant scaling up of the renovation rate and depth would add up to €90 billion per year.*

More details on the BPIE proposals (www.bpie.eu) were provided in 2020 in the document 'An Action Plan for the Renovation Wave', which provides guidance aimed at all major actors in the building sector. BPIE estimates that this program in the EU-27 will require €243 billion per year until 2050. Of this amount, €179 billion per annum is required for residential buildings, and €64 billion per annum for non-residential building renovations. It remains to be seen whether this massive sum will be made available. In either case, the outcome will provide valuable lessons to other jurisdictions with similar problems.

**Table 16.2.** Examples of targets and indicators from SDG Goal 13.

| Selected *Targets* from SDG 13 | Selected *Indicators* from SDG 13 |
|---|---|
| 13.1 Strengthen resilience and adaptive capacity to climate-related hazards and natural disasters in all countries | 13.1.1 Number of deaths, missing persons, and directly affected persons attributed to disasters per 100,000 population |
| | 13.1.2 Number of countries that adopt and implement national disaster risk reduction strategies in line with the Sendai Framework for Disaster Risk Reduction 2015–2030 |
| | 13.1.3 Proportion of local governments that adopt and implement local disaster risk reduction strategies in line with national disaster risk reduction strategies |
| 13.2 Integrate climate change measures into national policies, strategies, and planning | 13.2.1 Number of countries with nationally determined contributions, long-term strategies, national adaptation plans and adaptation communications, as reported to the secretariat of the United Nations Framework Convention on Climate Change |
| | 13.2.2 Total greenhouse gas emissions per year |

## 16.3 Climate change

Primary effects of high levels of atmospheric greenhouse gases (GHGs) include increased ambient temperatures, increased storm activity, variability in precipitation, riverine flooding, more variability in weather regimes, and increased sea levels.

Secondary effects include physical damage to buildings and infrastructure, droughts, and urban wildfires.

Tertiary effects include disruption to commercial links, reduced agricultural output, ecological damage, massive migrations of population, movements of wildlife and insects, and negative impacts on human productivity and health.

Addressing the multiple causes and impacts of climate change in more detail is a very complex task.

One helpful method for reducing confusion is to use the set of 17 interlinked sustainable development goals (SDGs) developed for the United Nations and adopted by the UN in 2015.

The SDGs are not useful at the level of Goals, but do become meaningful and useful at the subsidiary levels of Targets and Indicators, although still not very specific.

In table 16.2, we show a few examples of Targets and Indicators (UN report A/RES/71/313) from SDG Goal 13, *Take urgent action to combat climate change and its impacts.*

### 16.3.1 The drivers

As has been well established, the main driver of climate change is greenhouse gas emissions (GHGs) from anthropogenic sources such as transportation, electrical

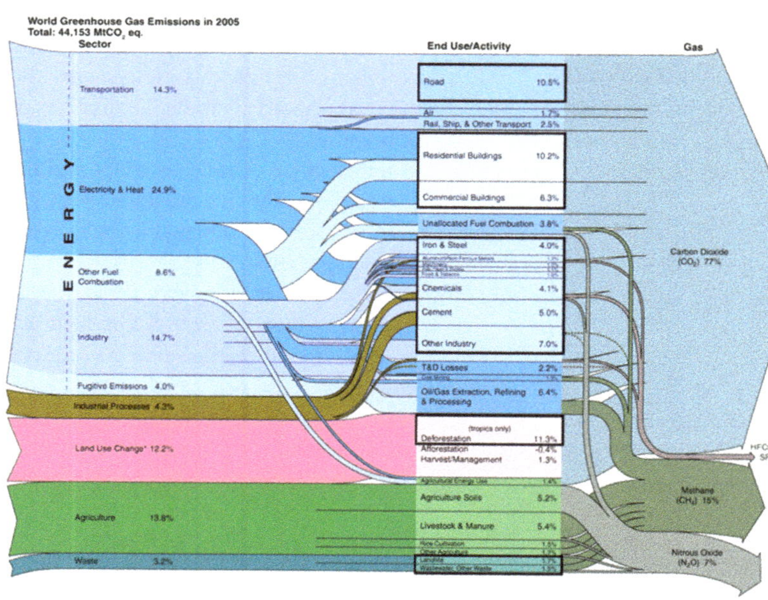


**Figure 16.3.** Energy inputs, end use activities, and greenhouse gas emissions, with construction-related activities highlighted (adapted from www.wri.org).

generation, fossil fuel combustion, industrial processes, land-use change, and agriculture. As the excellent graphic from the World Resources Institute (www.wri.org) (figure 16.3) indicates, energy inputs flow through many end uses or activities related to agricultural, industrial, transportation, and construction activities, producing GHGs such as carbon dioxide, methane, and nitrous oxide. The original diagram has been modified for our purposes (black boxes) to indicate the many activities that appear to be related to construction activities. It should be noted that the WRI diagram dates from 2008, but the general relationships remain the same.

From the United Nations Environment Programme Emissions Gap Report 2021 (UNEP 2021):

*Few of the G20 members' NDC targets put emissions on a clear path towards net-zero pledges. There is an urgent need to back these pledges up with near-term targets and actions that give confidence that net-zero emissions can ultimately be achieved and the remaining carbon budget kept.* (Page 20.)

*The emissions gap remains large compared to previous unconditional NDCs (Nationally Determined Contributions), the new pledges for 2030 reduce projected 2030 emissions by only 7.5 percent, whereas 30 percent is needed for 2 °C and 55 percent is needed for 1.5 °C.* (Page 23.)

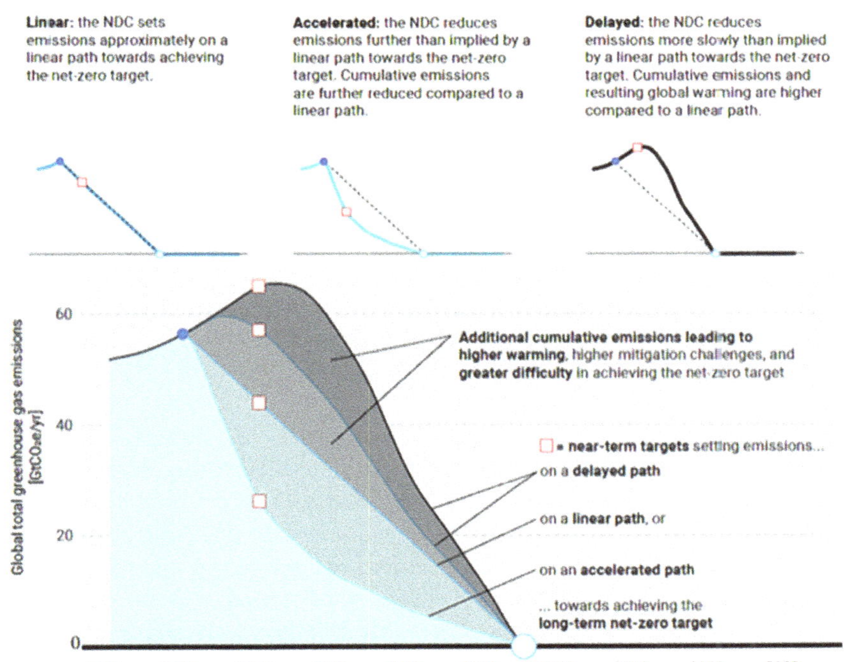


**Figure 16.4.** Annual emission reductions needed by 2025 to limit temperature rise to a maximum of 1.5 °C relative to 1990 (UNEP 2021).

### 16.3.2 Climate change impacts

The issues cited below outline a few examples of the most important types of impact, primarily related to urban areas.

*a. General impacts*

- Higher temperatures increase evaporation rates and aggravate potable and agricultural water shortages.
- Drought and soil shrinkage may cause structural damage to buildings.
- Ecological systems and the flora and fauna they support can be damaged.
- Insect migration will change under new temperature regimes.
- Wildfires may result in fringe urban areas during periods of drought, as occurred in 2016 in Fort McMurray in Alberta. Up to 88,000 people were forced from their homes, about 2400 buildings were destroyed, circa 590,000 hectares of forest were burned, and the total damage cost was approximately $9.9 billion CAD.
- In sub-arctic or arctic regions, high temperatures lead to the melting of permafrost, causing a release of methane (inducing more warming) and structural damage because of soil instability.
- In buildings, higher temperatures lead to an increase in the use of mechanical cooling, which creates more demand for electricity, which in turn creates more GHG emissions. Under such conditions, a traditional passive night cooling approach is not possible.

*b. An example: the European heatwave in 2003*

- Increasing ambient temperatures can drive many other changes. The 2003 heatwave in Europe demonstrated that heatwaves can cause an increase in death rates, especially in the older population (https://earthobservatory.nasa.gov/images/3714/european-heat-wave).
- The estimated extra mortality in eight European countries from the 2003 heatwave (Salagnac 2007) was 34,897.
- Electricity demand rose significantly because of the intense use of cooling systems, hydroelectric production was reduced by 19% because of reduced river flow rates, and nuclear production was reduced by 4% because the temperatures of river water used for cooling rose above acceptable levels.

*c. Impacts on human health and wellbeing*

A recent paper links climate change to a variety of human health impacts (Kjellstrom *et al* 2016). The following text is adapted from the abstract:

*Ambient heat exposure is a well-known health hazard which reduces human performance and work capacity at heat levels already common in tropical and subtropical areas. Increasing heat exposure during the hottest seasons of each year is a key feature of global climate change; heat exhaustion and reduced human performance are often overlooked in climate change health-impact analyses. Later this century, many among the four billion people who live in hot areas worldwide will experience significantly reduced work capacity owing to climate change. In some areas, 30%–40% of annual daylight hours will become too hot for work to be carried out. The social and economic impacts will be considerable, with global gross domestic product (GDP) losses of more than 20% by 2100.*

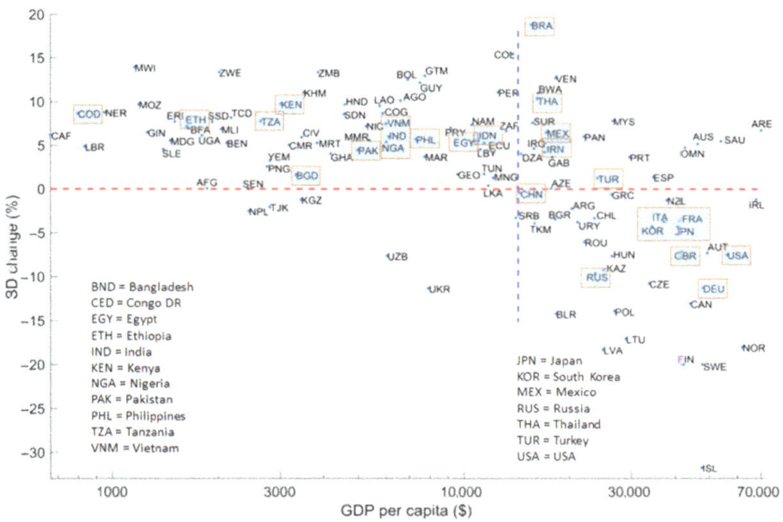


**Figure 16.5.** Relative change in the standard deviations (SD) of monthly temperature anomalies between 1990 and 2013 (adapted from Bathiany *et al* 2018).

*d. Temperature variability v. GDP*

A recent study *Climate models predict increasing temperature variability in poor countries* (Bathiany *et al* 2018) has provided some alarming conclusions about the locations and GDPs of countries that will be most exposed to climate change impacts. According to the authors:

*We show that climate models consistently project increases in temperature variability in tropical countries over the coming decades, with the Amazon as a particular hotspot of concern…. Mechanisms include drying soils and shifts in atmospheric structure. Outside the tropics, temperature variability is projected to decrease on average because of a reduced meridional temperature gradient and sea-ice loss…. The countries that have contributed least to climate change, and are most vulnerable to extreme events, are projected to experience the strongest increase in variability. These changes would therefore amplify the inequality associated with the impacts of a changing climate.*

Figure 16.5 is adapted from the paper cited above and shows the relative change in the standard deviations (SD) of monthly temperature anomalies up to the end of the 21st century versus per-capita GDP between 1990 and 2013 in different countries. The red line marks zero change in the (predicted) temperature variability. The blue lines mark half of the world population. Changes in figure 16.5 include highlighting countries of more than 50 million.

*e. Sea-level rise*

Sea-level rise is an important factor in assessing the security of populations living in vulnerable areas; vulnerable populations number in the hundreds of millions. One of the principal research organizations working in this area, Climate

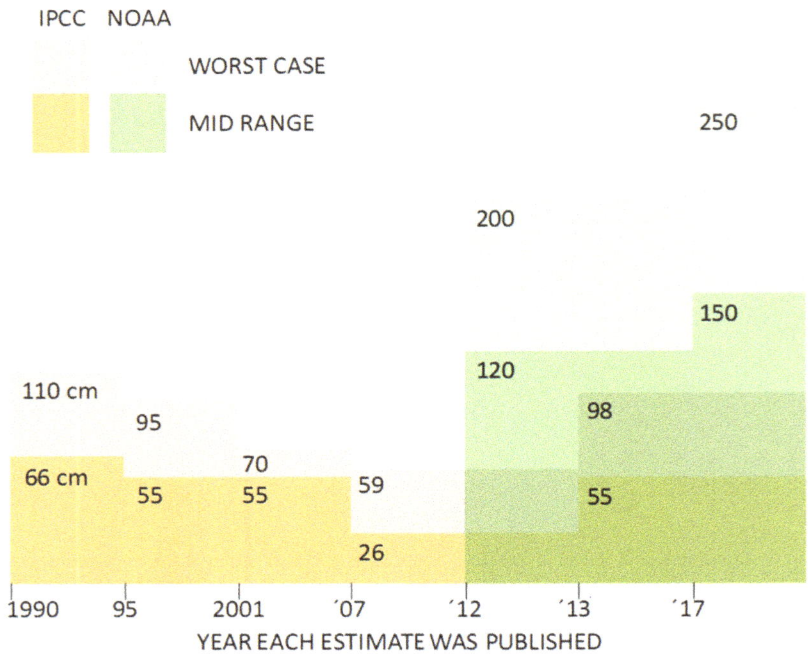


**Figure 16.6.** Changing estimates of sea-level rise by 2100 (data from Lindsey 2021).

Central, stated in a recent paper (Kulp and Strauss 2019): *the new figures are… revealing that coastal elevations are significantly lower than previously understood across wide areas.*

The conclusions drawn from this work are that the populations living below the average annual flood level in 2050 will include 93 million people in China, 42 million in Bangladesh, 36 million in India, 31 million in Vietnam, 23 million in Indonesia and 12 million in Thailand, with a total of 237 million in Asia, and 300 million globally. It should be noted that these projections are based on assumptions of moderate emissions cuts (RCP 4.5), and do not factor in potential coastal defences, such as seawalls or levees.

Because many large cities were established in coastal areas, major industrial and infrastructure installations are also vulnerable. It is clear that the rise in sea level will have major physical, economic, and societal impacts by 2050, and that massive migrations of climate refugees will be a likely result.

Figure 16.6 is based on data from Climate Change Central (Lindsey 2021) and shows various projections of sea-level rise according to the year the estimates were published.

## 16.4 Integrated action

A logical action plan to respect the needs of both post-pandemic recovery and climate action will include the following measures. Based on current trends and climate change impacts to date, the following actions, regionally modified, are needed if we are to be ready to withstand the next pandemic and to minimize any increase in ambient temperatures above 1.5 °C.

### 16.4.1 Urgent action

1. Protect critical facilities and infrastructure (harbors, airports, communication networks, water and waste water systems, urban public transit, major roads, emergency services, hospitals, schools, etc.) from climate change impacts.
2. Take immediate steps to rebuild the confidence of users in urban public transit systems.
3. Provide short-term accommodation for population groups forced to move by climate change impacts or pandemic isolation requirements.

### 16.4.2 Immediate organizational and administrative steps

1. Integrate climate and post-pandemic actions so that post-pandemic social and economic goals and 2050 zero GHG targets can be met. A bottom-up approach that involves local groups may prove to be the most effective.
2. Designate redevelopment authorities at national and local levels to establish or modify relevant regulations and to provide loans and grants for redevelopment to meet post-pandemic rebuilding needs and climate change adaptation and/or mitigation needs.
3. Use the UN SDGs as a guide to identify the most relevant issue areas for action. SDG 11 (*Sustainable Cities and Communities*) and SDG 13 (*Take urgent action to combat climate change and its impacts*) *are* the key goals for the issues covered in this document (https://sdgs.un.org/goals).
4. Undertake urban area audits, including all issues and systems relevant to preparing for the next pandemic and for anticipated climate change impacts. The relevant items include the mitigation of urban heat islands, land use, ecological systems, above- and below-grade infrastructure, public transport systems, and buildings.
5. Establish performance targets for the major urban systems audited in (7), aiming for key dates such as 2025, 2030 and 2050.
6. Ensure that estimates of aggregate lifecycle carbon emissions for new construction are included in Nationally Determined Contributions (NDC) that are prepared by national governments.
7. Obtain funding commitments for urban infrastructure and housing renovation programs from financial institutions and identify major private sources of capital.
8. Limit peak electrical demand through incentives and regulations to minimize the need for new generating capacity.
9. Introduce incentives to rapidly move towards zero use of fossil fuels and to promote clean energy and renewables. An effective mechanism for achieving this is a carbon tax (WBG 2019).

10. Support a shift from private vehicles to public transport to reduce the waste of urban land, reduce the consumption of fossil fuels, and improve urban air quality.
11. *Freeze* new construction activity in areas that are vulnerable to climate impacts or are of high ecological value.
12. Limit new construction in other areas that have potential value for agriculture.
13. Adjust taxation systems and building regulations to:
    - provide incentives to maintain and retain heritage buildings.
    - provide incentives for renovation and disincentives for new construction.
    - constrain the provision of dwelling sizes or levels of service that greatly exceed functional requirements.
    - incentivize the efficient use of non-primary dwellings.
14. A significant proportion of building types (e.g. office or retail commercial), or parts of buildings, may now be surplus to requirements because of the pandemic, and should be seen as an important resource to serve new post-pandemic and climate action needs.
15. Introduce incentives to reduce waste in design, construction, or operation (total energy consumption and emissions caused by secondary houses, excess floor area, equipment, appliances, or building services that exceed functional necessities).
16. Require that submissions for building permits should include estimates of embodied GHG emissions for all permanent elements of the building.
17. Establish requirements for tri-annual operating permits for new buildings, including very demanding targets for nearly zero operating energy consumption and GHG emissions.
18. Establish public building performance databases so that progress can be measured and compared.
19. Use local labour to the maximum extent possible and establish training programs to support all the initiatives outlined in this section.

### 16.4.3 Planning and program initiatives

1. Update or prepare neighbourhood plans to maximize pedestrian, bicycle, and other non-vehicular mobility using the 'Paris 15-minute city' approach (Yeung 2021)...*The concept is to improve quality of life by creating cities where everything a resident needs can be reached within a quarter of an hour by foot or bike. The 15-minute city requires minimal travel among housing, offices, restaurants, parks, hospitals and cultural venues. Each neighborhood should fulfil six social functions: living, working, supplying, caring, learning and enjoying*....
2. Allocate surplus and vacant urban fringe land for aquifer replenishment, urban forests, wildlife, and biodiversity.
3. Encourage urban agriculture for local residents to reduce transport emissions and to increase neighbourhood access to fresh produce.

4. Measures may be required to control the growth of undesirable insect populations that may be enabled by the increasing ambient temperatures.
5. Introduce high-efficiency public transport systems, with stops located at convenient and strategically important locations to increase usage and to convince users to leave cars at home.
6. Mitigate the urban heat-island effect, reduce the energy demands made by buildings, and increase the livability of the urban area by increasing effective areas of vegetation, parks, and urban forests and by (in the long run) adjusting street design. All of this is highly dependent on regional characteristics (https://www.epa.gov/heatislands).
   - Urban forests can provide shade and evaporative cooling during warm seasons. Coniferous trees can reduce undesired wind speeds during cold periods. Open areas can be used as skating rinks in winter.
   - Shading by trees and other vegetation is important for health and comfort and to reduce cooling loads.
   - Evapo-transpiration from deciduous trees adds a cooling effect, but in tropical and subtropical regions high ambient temperatures and relative humidity limits comfort.
   - Green building elements, such as green roofs and walls can mitigate the urban heat-island effect, and simultaneously contribute to rainwater management (even during heavy storms), air quality (through the sequestration of pollutants and dust), and improvement in thermal and acoustic isolation.
   - Walking and bicycling are important strategies in all seasons in most regions. Shaded and dedicated routes for walking and bicycling are important strategies.
   - Car use and parking will remain important in areas with cold winters. An important issue is how to use traffic changes or incentives to effectively switch the emphasis to bicycling during warm seasons.
   - Increase the reflectivity (albedo) and permeability of paved surfaces in streets and public areas.
   - Narrow streets allow buildings to shade each other during the day, to minimize solar gain or glare.
7. Provide storm water management systems to accommodate the anticipated variability in precipitation, especially more frequent occurrences of sudden storms or flash floods.
   - Permeable soil will allow precipitation to replenish aquifers.
   - In areas where water conservation is a priority, measures to reduce water consumption can include the reuse of storm water via holding ponds, swales, underground storage tanks, green roofs, and bioretention areas.
   - Soil cells that capture storm water runoff promote the healthy growth of trees (STEP 2020) and bioretention areas can reduce pollutant loads in watercourses and groundwater.

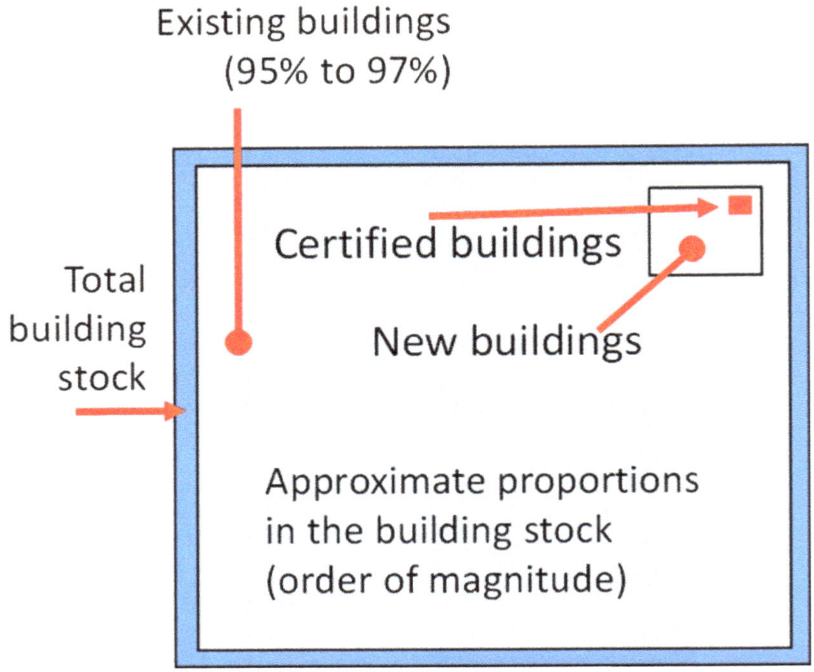


**Figure 16.7.** Order-of-magnitude proportions of total stock, existing buildings, new buildings, and certified buildings.

8. Make use of the urban areas and buildings made redundant by the pandemic:
   - Convert unused open parking spaces to parks, tree cover, or open-air recreation facilities.
   - Convert surplus office building space for residential, educational, or public uses.
   - Convert other surplus buildings, e.g. conference centres, gymnasia, shopping centres, parking garages, or parts of airports to new uses.
   - Establish refuge centres in neighbourhood schools, community centres, or similar facilities, for use in health and climate emergencies.
9. The stock of existing buildings normally constitutes a large (94% to 97%) proportion of the total building stock, and if existing buildings can be retained as a viable component of the total stock, much new construction and the related greenhouse gas emissions can be avoided (figure 16.7).

The preservation of existing buildings must therefore be a priority, and this can best be done by launching 'deep green' renovation programs that will upgrade the pandemic and climate performance of existing buildings.

1. Launch programs for affordable housing aimed at low-income populations as a major element of post-pandemic social equity and economic recovery measures.

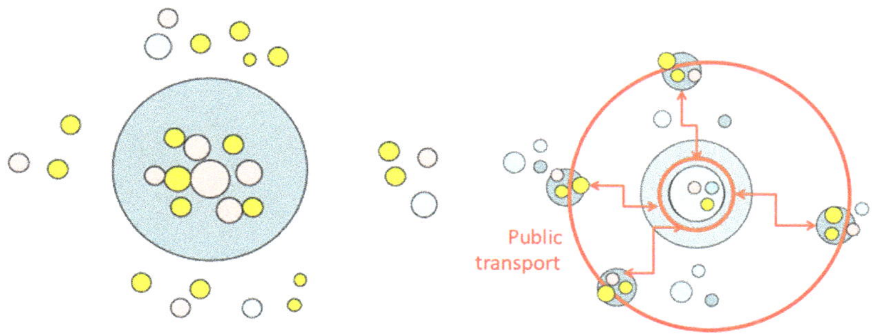


**Figure 16.8.** Urban scatter (left) and satellite communities connected by public transit (right).

2. Support commercial retail businesses and ensure the preservation of the existing commercial building stock through renovation.
3. In developed western countries, the pandemic has led to a major increase in the demand for single-family houses with gardens in suburbs and small towns, changing the traditional pattern of high densities and mixed uses in urban centres (figure 16.8 left). Climatic as well as pandemic needs will be supported if planners aim for high-density mixed-use satellite nodes connected by public transport (figure 16.8 right). This type of model is not particularly appropriate or achievable in developing countries.

### 16.4.4 Building design and technology initiatives

1. Adjust occupant populations in all non-residential buildings to support increased distancing and limits on occupant density.
2. In new buildings, specify a high degree of flexibility of internal systems, to facilitate changes in function.[2]
3. Provide all new dwelling units with private outdoor space, e.g. gardens, courts, or balconies. In regions with sub-zero winter conditions, balconies should be designed with thermal breaks.
4. Minimize embodied energy and lifecycle emissions in new construction. One effective technique is to retain existing buildings and their structures and to provide additions for new marginal requirements.
5. Inter-building performance synergies are possible in clusters of buildings that have varied functions and configurations, by using storage and control systems to share surpluses and deficits in thermal energy, renewable power generation, and grey water. This concept is based on the fact that various building and occupancy types have different requirements for the consumption of

[2] It is easier to convert an office or public building to residential use than the reverse, because of the typically greater floor loading capacities and ceiling heights of public and office building types.

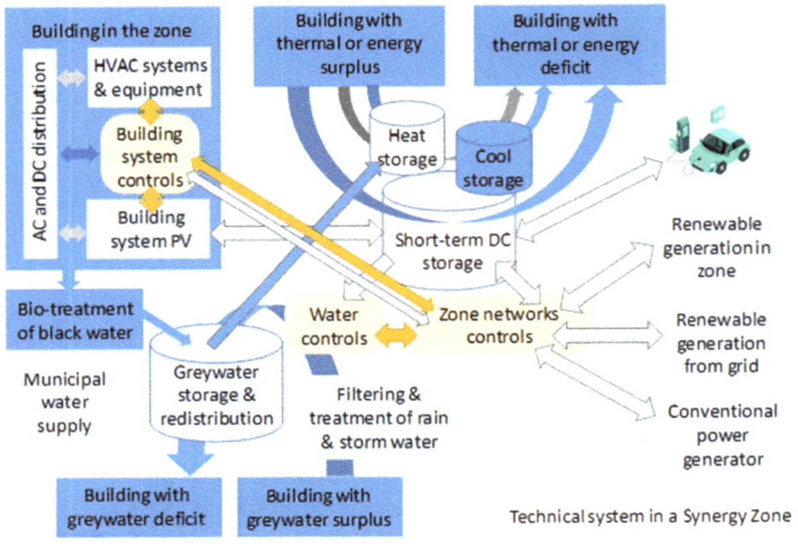


(a)

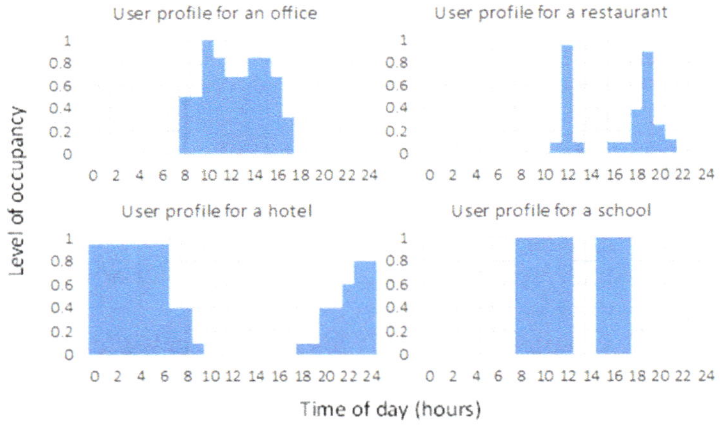


(b)

**Figure 16.9.** Opportunities for synergies from: (a) different occupancies and building types (by the author), and (b) diversity in occupancy profiles.

thermal energy, power, and water, and also have different potentials to satisfy these demands (Larsson *et al* 2013) (figure 16.9(a)). Variations in peak usage hours for various occupancy types also enable the load sharing of thermal energy, power, and grey water, assuming that appropriate agreements, storage, and control systems are in place (BS EN 15232–1:2017) (figure 16.9(b)).

1. The target for operating energy consumption for all new buildings should be zero or nearly zero, with a maximum use of renewable energy.
2. Upgrade ventilation systems in long-term care facilities, schools, and office buildings as per (42) and (43) to reduce the risk of COVID-19 infection.
3. In new and existing buildings, encourage natural and hybrid ventilation, increase standards for air changes per hour (ACH), and increase outdoor air ratios in ventilation systems. In cold climates, this is a rare example of a conflict between pandemic and climate action needs, but the problem can be minimized by the use of heat-recovery systems.
4. Operable windows provide the simplest and most direct source of outdoor air, but in regions with cold winters they tend to remain closed to avoid energy loss and cold drafts. Under such conditions, the locations of opening window segments are important to maintain comfort conditions.
5. Ensure that the energy consumption, energy efficiency, GHG emissions, water consumption, and wastes during a building's operation conform to design performance specifications. This can be ensured through the application of a post-occupancy operating audit carried out two years after occupancy.
6. Augment normal lift capacities to ensure reduced passenger densities of approximately 3 $m^2$ or more per passenger. In some cases, this will not be physically possible, and alternate strategies for reduced service levels and/or scheduled access will be required.
7. In large buildings with public access, establish building entry checkpoints, provide scanning and monitoring of occupants' health status and movements within the building.
8. Strengthen equipment and appliance efficiencies to at least A++ (U.S. or Canada) or the equivalents (in the EU, this will be 'A' as from 2021).
9. Monitor waste water flows for COVID-19 and other viruses at the level of urban blocks or larger buildings. The intermittent monitoring and analysis of untreated sewage for SARS-CoV-2 residue gives an indication of the presence of COVID-19 in the community (Orive *et al* 2020).

## 16.5 Conclusions

- An effort to simultaneously deal with post-COVID-19 and climate action issues appears to be necessary and feasible.
- Such efforts will offer great opportunities to integrate post-pandemic goals with those of climate action.
- Figure 16.10 identifies the relevance of each action to pandemic and climate change requirements. It can be seen that 25 of 48 total actions are applicable to both pandemic and climate change requirements, and 2 actions conflict. This indicates that attempts to combine actions to address both of these very major issues are worthwhile and should be pursued vigorously. However, specific actions must take into account the regional context.

| Actions | Covid-19 Pandemic | Climate Change |
|---|---|---|
| Urgent actions 1 to 3 | These three actions are applicable to both requirements | |
| Immediate organizational and administrative steps 4 to 22 | Actions 4-8, 10, 17, 21 and 22 are directly applicable to pandemic requirements | Actions 4-16 and 18-22 are directly applicable to climate change requirements |
| | 9 of 18 actions are applicable to both requirements | |
| Planning and program initiatives 23 to 34 | Actions 23, 25, 28 and 30-34 are directly applicable to pandemic requirements | Actions 23-33 are directly applicable to climate change requirements |
| | 10 of 11 actions are applicable to both requirements | |
| Building design and technology initiatives 35 to 48 | Actions 35-37, 41-46 and 48 are directly applicable to pandemic requirements | Actions 36, 38-40, 42, 44 and 47 are directly applicable to climate change requirements.<br>Actions 35 and 41 conflict with climate change requirements. |
| | 3 of 13 actions are applicable to both requirements<br>2 actions conflict | |
| 25 of 48 total actions are applicable to both pandemic and climate change requirements, and 2 actions conflict | | |

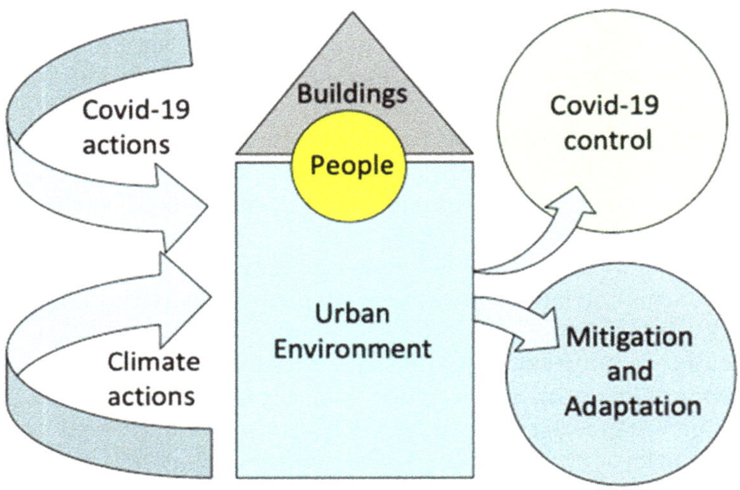


**Figure 16.10.** Linking integrated action items to the two crises.

## Acknowledgments

The author acknowledges and is grateful for contributions by Katerina Tsikaloudaki, Richard Lorch, and Brian Hierlihy.

## Bibliography

ASHRAE 2020 *ASHRAE Position Document on Infectious Aerosols*, Atlanta. https://ashrae.org/file%20library/about/position%20documents/pd_infectiousaerosols_2020.pdf

Bathiany S, Dakos V, Scheffer M and Lenton T M 2018 Climate models predict increasing temperature variability in poor countries *Sci. Adv.* **4** eaar5809

Hepburn C, O'Callaghan B, Stern N, Stiglitz J and Zenghelis D 2020 *Will COVID-19 fiscal recovery packages accelerate or retard progress on climate change?* Smith School Working paper 20-02

Kjellstrom T, Briggs D, Freyberg C, Lemke B, Otto M and Hyatt O 2016 Heat, human performance and occupational health: a key issue for the assessment of global climate change impacts *Annu. Rev. Public Health* **37** 97–112

Kulp S A and Strauss B H 2019 New elevation data triple estimates of global vulnerability to sea-level rise and coastal flooding *Nat. Commun.* **10** 4844

Larsson N, Salat S and Bourdic L 2013 Performance synergies in small urban zones *iiSBE* (https://researchgate.net/publication/259851435_Performance_Synergies_in_Small_Urban_Zones)

Lindsey R 2021 Climate change: global sea level (https://climate.gov/news-features/understanding-climate/climate-change-global-sea-level)

Mohan S V, Hemalatha M, Kopperi H, Ranjith I and Kumar A K 2021 SARS-CoV-2 in environmental perspective: occurrence, persistence, surveillance, inactivation and challenges *Chem. Eng. J.* **405** 126893

Orive G, Lertxundi U and Barcelo D 2020 Early SARS-CoV-2 outbreak detection by sewage-based epidemiology *Sci. Total Environ.* **732** 139298

UNEP 2021 (United Nations Environment Programme) 2021 Emissions Gap Report 2021 https://www.unep.org/resources/emissions-gap-report-2021

Salagnac J-L 2007 Lessons from the 2003 heat wave: a French perspective *Build. Res. Informat.* **35** 450–7

STEP 2020 Assessing the Health of Toronto Street Trees Irrigated by Storm Water Technical Brief, Prepared by TRCA

WBG 2019 World Bank Group, Carbon Pricing Leadership Coalition Report of the High-Level Commission on Carbon Pricing and Competitiveness (https://elibrary.worldbank.org/doi/abs/10.1596/32419)

Yeung P 2021 A new urban planning model will change the French capital—and could provide a template for how to create stronger local communities and make residents happier *BBC* www.bbc.com/worklife/article/20201214-how-15-minute-cities-will-change-the-way-we-socialise)

www.ingramcontent.com/pod-product-compliance
Ingram Content Group UK Ltd.
Pitfield, Milton Keynes, MK11 3LW, UK
UKHW051900100926
8556IPUK00005B/704

9 780750 339698